An Introduction to Design Arguments

The history of design arguments stretches back to before Aquinas, who claimed that things which lack intelligence nevertheless act for an end to achieve the best result. Although science has advanced to discredit this claim, it remains true that many biological systems display remarkable adaptations of means to ends. Versions of design arguments have persisted over the centuries and have culminated in theories that propose an intelligent designer of the universe. This volume is the only comprehensive survey of 2,000 years of debate, drawing on both historical and modern literature to identify, clarify, and assess critically the many forms of design argument for the existence of God. It provides a neutral, informative account of the topic from antiquity to Darwin, and includes concise primers on probability and cosmology. It will be of great value to upper-level undergraduates and graduates in philosophy of religion, theology, and philosophy of science.

BENJAMIN C. JANTZEN is Assistant Professor of Philosophy at Virginia Tech. He has published papers in journals including *Proceedings of the National Academy of Sciences*, *Biophysical Journal*, *Philosophy of Science*, and *Synthese*. His work on formal methods in the philosophy of religion has been published in *Transactions of the Charles S. Peirce Society* and in the anthology *Probability in the Philosophy of Religion* (2012).

An Introduction to Design Arguments

BENJAMIN C. JANTZEN

Virginia Tech

Shaftesbury Road, Cambridge CB2 8EA, United Kingdom

One Liberty Plaza, 20th Floor, New York, NY 10006, USA

477 Williamstown Road, Port Melbourne, VIC 3207, Australia

314–321, 3rd Floor, Plot 3, Splendor Forum, Jasola District Centre, New Delhi – 110025, India

103 Penang Road, #05–06/07, Visioncrest Commercial, Singapore 238467

Cambridge University Press is part of Cambridge University Press & Assessment, a department of the University of Cambridge.

We share the University's mission to contribute to society through the pursuit of education, learning and research at the highest international levels of excellence.

www.cambridge.org
Information on this title: www.cambridge.org/9780521183031

First published 2014

A catalogue record for this publication is available from the British Library

Library of Congress Cataloging-in-Publication data
Jantzen, Benjamin C., 1977–
An introduction to design arguments / Benjamin C. Jantzen.
    pages   cm
Includes bibliographical references and index.
ISBN 978-1-107-00534-1 (hardback) – ISBN 978-0-521-18303-1 (pbk.)
1. Teleology.   I. Title.
BD531.J36 2014
124–dc23
2013036672

ISBN   978-1-107-00534-1   Hardback
ISBN   978-0-521-18303-1   Paperback

To Parisa

Contents

Figures and tables

Figures

Tables

Preface

This book is a critical survey of *design arguments*, attempts to infer the existence of a God or gods by demonstrating the likely role of intelligence in shaping the world of experience. By critical, I do not mean polemical. What follows is not an attempt to dismiss design arguments as categorically misguided or ill-conceived. Nor is it a religious apologetic. Rather, it is intended to be a neutral philosophical reconstruction and analysis of the entire field of design arguments advanced from the rise of Western philosophy in ancient Greece to the present day.

The treatment of this material is introductory in a couple of ways. As with any introduction to a field of study, I have sacrificed some depth in favor of breadth in order to give the reader a coherent picture of the entire landscape of the design debate. My aim is in part to provide a synopsis of a long philosophical conversation so that someone interested in working on the philosophical puzzles surrounding design arguments can jump right in. The book does not assume that the reader is familiar with the jargon of academic philosophy or has had any formal training in the analysis of arguments, and so can serve as the basis for an introductory course in philosophy or critical thinking.

There are, however, a number of ways in which this book goes beyond a mere introduction to its subject. To begin with, no one has attempted a comprehensive survey of design arguments since L. E. Hicks published his wonderful book, *A Critique of Design-Arguments*, in the late nineteenth century. While I do not pretend to have a written a book as clever and engaging as Hicks', the survey that follows may nonetheless serve as a guide and reference for scholars actively working in the field. Furthermore, it is my hope that such breadth has not come at the cost of philosophical rigor. I offer many critiques and analyses that do not appear elsewhere. These include interpretations of the work of William Paley in Chapter 8, a review of modern concepts of complexity in Chapter 14, a rebuttal of the objection from 'Observation Selection Effects' in Chapter 18, and an assessment of the metaphysical assumptions implicit in fine-tuning arguments, also in Chapter 18.

The intended audience for this book is broad and varied. The book emerged from an undergraduate course I developed at Carnegie Mellon University. The course offered undergraduates an introduction to philosophical reasoning and argument analysis in the context of the pursuit of a single question: can we infer the existence of God from empirical evidence of intelligent agency? Aside from addressing a neglected philosophical issue of deep importance to many, such a course offered the best of both worlds for an introductory course: the chance to engage deeply with a single complex issue and, at the same time, to gain experience with many different argument strategies. After all, for virtually every argument form that bears a name, someone has offered a corresponding design argument. This book is the textbook I wish I had available for that course. Thus, my intended audience includes undergraduate students and the instructors who may wish to offer such a course. At the same time, each chapter, particularly those concerning modern arguments, is a review of the current state of the art. As such, graduate students and professional philosophers should find the discussion and the list of references sufficiently rich to guide further research into the topic. Finally and perhaps most importantly, I offer this book to the general public. Interest in the question of whether one can establish the existence of a deity on the basis of experience extends well beyond the academy and those for whom philosophy is a vocation. I have tried to provide enough background information to make this text accessible to anyone with such an interest.

As a guide to the reader or course instructor, I'd like to point out some features of the book's structure. The text divides into two parts. The first ten chapters are dedicated to searching out design arguments in the historical record from antiquity through the introduction of Darwin's theory of natural selection in the mid nineteenth century. There was a more or less continuous philosophical debate over design spanning this period, and the aim of the first part of the book is to carefully reconstruct from this conversation as many types of design argument as possible, catalogue the criticisms and rebuttals offered by historical figures, and assess each type of argument in light of modern empirical knowledge. I want to make the strongest case possible for design, and this means exploring all available options. To determine what these options are, the historical record is treated as a great repository of possible argument strategies. The second half of the book examines closely those design arguments that have been advanced and debated in the modern literature. These arguments make frequent and essential use of formal tools such as probability theory, and appeal to the details of complex contemporary physical theories. For these reasons, the material in this portion of the book, though I present it without assuming any particular expertise on the part of the reader, is generally more challenging.

There are quite a few people from whom I and this book have benefited greatly, and they deserve to be recognized. I would first like to thank the Philosophy Department at Carnegie Mellon University (CMU), especially Richard Scheines and Mara Harrell, for giving me the opportunity and resources to develop and implement the undergraduate course out of which this book ultimately developed. And, of course, I am grateful to the students who participated in "Life, the Universe, and God" on the two occasions it was offered at CMU (in the spring of 2009 and 2011). Their enthusiasm, searching questions, and criticisms played a substantial role in shaping this book. I am also indebted to Walter Ott, Clark Glymour, the students of "God in the West" (CMU, fall 2013), Robert Camp, Kristina Jantzen, Ronald Jantzen, and an anonymous reviewer for Cambridge University Press for helpful discussion and critical comments on portions of the manuscript. Timothy Graham is due special thanks for his assistance in indexing this text. I am grateful as well for the editorial efforts of Hilary Gaskin. In addition to the essential role she played in getting this book published, her suggestions about structure and tone were astute, and made this book much better than it might have been. Thanks also to Anna Lowe, especially for her patience with my frequent delays in turning over materials. The input from all of these people has shaped this book in important ways for the better, and all of the errors and defects of fact, reason, or judgment that remain are, of course, solely my responsibility.

With regard to the role others have played in producing this book, I would like to emphasize the special contribution that Robert Camp has made. His original black-and-white illustrations are scattered throughout the chapters that follow. Their value is threefold. Pragmatically speaking, they are essential tools for clarifying concisely some of the difficult concepts encountered in this text. His drawing of the Antikythera Mechanism, for instance, is surely worth a thousand words. Second, these beautifully rendered drawings grace the text with an aesthetic value that is rare in modern academic work. They are a joy to look at. Finally, the drawings provide a visceral connection to a time when natural history and natural science were illustrated with the pen, not the computer. The practice of natural history illustration is deeply interwoven with the rise and development of natural science and of design arguments.

Finally, I offer my most profound gratitude to my wife, Dr. Parisa Farhi. Without her support, neither the original university course on design arguments nor this book would have been possible.

1   Introduction

1.1 What this book is about

Broadly speaking, we human beings use two conceptual schemes or 'paradigms' to explain the world in which we find ourselves.[1] In the 'teleological paradigm' natural events are explained in terms of the same sorts of purposes, means, and goals we use to explain our own behavior. Within the 'naturalistic paradigm', the world is explained in terms of natural laws and mindless processes. The naturalistic paradigm has been enormously successful. As an underlying framework for science, this way of approaching the world has yielded a vast knowledge of natural phenomena, as well as the technology which distinguishes our modern way of life from all that came before. The teleological paradigm on the other hand is ancient and deeply intuitive. The oldest human accounts of nature were made from this perspective. The creation stories of cultures around the world – from the Babylonian account in which Merodach fashions the world from the corpse of the great Mother dragon Tiamat,[2] to the biblical story of Genesis – are examples of teleological explanations. While such explanations have been largely displaced by appeals to natural law, one product of the teleological paradigm continues to remain relevant: the family of 'design arguments' for the existence of God.

Design arguments are characterized, not surprisingly, by appeals to design. Each such argument urges us to accept that one or another aspect of the world or of things in the world is the product of purposeful, intelligent agency. That is, each design argument attempts to establish that some aspect of the natural world was designed. From there, it is a short

---

[1] A graceful argument to this effect was made by the pioneering psychologist C. Lloyd Morgan (1906) in his Lowell lectures.

[2] MacKenzie 1915, 138–62.

mental hop to the existence of a designer. After all, the presence of design in the world surely implies the existence of a designer. For the arguments we will examine, that designer is typically (though not always) understood to be the Christian God.

This book is a systematic attempt to determine which, if any, design arguments are convincing. Our goal is to ascertain whether any such argument succeeds – or has the potential to succeed – in establishing the existence of a god or gods on the basis of our experience of the world. To do so, we will need to survey the available arguments, isolate the essential form of the inferences made, and assess the merits of each type of inference. This book is organized into two parts. In the first part, we'll consider the major kinds of design argument from antiquity through the mid nineteenth century CE. We will attempt to classify each of these arguments with respect to its logical structure, and consider the strengths and weaknesses of each, particularly in the light of what various historical critics have had to say. In the second part, we'll turn to design arguments in the modern literature, each of which is scrutinized in detail. There are two major classes of argument to consider here: biological and cosmological. In the former class are arguments that infer an intelligent designer from the properties of organisms, or from the apparently purposeful arrangement of the parts of organisms. In the cosmological class, the argumentative focus is on the appearance of 'cosmic fine tuning' – if the values of many physical constants had been even slightly different from what they actually are, life would have been impossible. In various ways, this apparent coincidence is used to argue that the universe as a whole must have been designed.

Since our goal is to assess the merits of each argument in its strongest form, we will have to pay special attention to the structure of arguments and the ways in which some claims about the world rationally compel belief in others. I introduce some useful tools for this purpose in Chapter 2. After describing the major kinds of inference – ways in which one set of claims can provide evidence in favor of another – I explain the method of argument diagramming, a systematic approach for representing and criticizing arguments that will be used throughout this text. Aside from the basics of argument analysis, some of the design arguments we will consider demand familiarity with particular mathematical or scientific topics. Whenever this is the case, I provide a primer for the non-specialist.

The philosophical question this book sets out to address – are there any sound design arguments for the existence of God? – has been the subject of one long conversation over the past 2,000 years. While it is possible to enter this conversation and this book at any point and still extract something of value, each argument is best assessed in the context of what has come before. Some approaches, like the argument from analogy which we will study in Chapter 7, were eventually abandoned after collapsing under criticism. Some newer arguments, such as Paley's version of the argument from order (Chapter 8), were structured in a particular way expressly to avoid the criticisms that doomed older approaches. Knowing this makes assessing newer arguments a much more efficient process. The chapters of this book are intended to be read sequentially, with each building upon the last. With each new argument in the series, new conceptual tools are introduced, fresh criticisms are added to a growing stockpile, particular argument forms are abandoned, and potentially fruitful suggestions are pursued. For this reason, I encourage the reader to get a sense of the overall conversation before jumping into any particular argument.

1.2 Intuitions of design

Teleological explanations are appealing. Without giving the matter much thought, many aspects of the natural world just seem as if they were arranged that way on purpose. These reflexive reactions to the world are strong enough to have maintained the appeal of design arguments for a very long time. So before we turn our attention to the design arguments themselves, it will help to have a sense of what features of the world provoke an attribution of design in the first place.

There are three principal kinds of intuition that drive interest in design arguments. By 'intuition' I mean a pre-reflective assertion or explanation – something like a 'gut reaction'. I'll refer to the three pertinent kinds of intuition as 'purpose', 'form', and 'conspiracy'. *Purpose* pertains to the way in which parts of the world – generally living things – relate to one another. In particular, such an intuition concerns the way in which things seem to be suited to a goal that does not originate with themselves. It is difficult not to speak of the parts or behaviors of an organism without referring to the tasks they were intended to perform. For instance, the wing of a bird has a peculiar shape, an extraordinary ratio of strength to weight, and a pronounced size relative to the bird, all of which appear to be for the purpose of flying. If asked why a bird has wings, the natural response is "So it can fly." Every 'adaptation' of an organism, every trait which suits a particular organism to survival in a particular environment, gives the impression of having been intentionally arranged for the express purpose of sustaining that kind of organism. Since the bird does not design its wings or the giraffe its neck, we look elsewhere for the author of these adaptations. Of course, we cannot ignore evolution by natural selection, always the elephant in the room when discussing apparent purpose in nature. In a scientifically literate society it is difficult to suppress the evolutionary response to intuitions of purpose in the structure of organisms, namely that these structures are the product of natural selection not intelligent design. We will consider Darwin's theory in Chapter 9, and the profound impact of his work on design arguments, particularly those that appeal to adaptation in organisms, in Chapters 12 and 13. But for now, try not to draw hasty conclusions. At this stage, I merely want to bring into focus the pre-reflective impressions that continue to make design arguments compelling.

To get an idea of what I mean by intuitions of *form*, think of pyramids and clocks. Intuitions concerning form involve an immediate recognition of properties that result only or mostly from acts of intentional design. These include symmetry, geometric simplicity, order, precision, and complexity. The properties of geometric simplicity and order are everywhere in human architecture (various modernist buildings notwithstanding). On the scale of everyday human experience – characterized by lengths on the order of 1 meter – only intentional design results in the production of rectangles, circles, parallelepipeds and the like. The simple geometric form of the pyramids at Giza, the elegant geometric ratios of the Parthenon, and the precisely level surfaces of an airport runway are all recognizable products of design. Instances of geometric elegance also occur in nature, though on a smaller scale. Each snowflake has a simple sixfold symmetry – rotate one 60° around an axis through its center and it looks the same as it did before. The question, of course, is whether this sort of natural geometric property is also the product of design.

Our intuitions of form are especially strong when geometric simplicity is combined with complexity. Clocks are the classic example. Clockwork is a common metaphor in Enlightenment discussions of design. Clocks and other complicated machines involve many parts, each of which displays the sort of geometric simplicity mentioned above. Furthermore, each part of a watch plays a very specific functional role, which it would fail to perform if its shape or other physical properties fell outside of a very narrow range. When every part performs its function, the result is a coordinated series of causal relations – a spring pushes a cog which turns another, which turns another, which turns the hour hand. Various parts of the universe or perhaps even the universe as a whole seem to exhibit this sort of complex interaction, an interaction that would devolve into chaos or freeze into a static lump if some of the interactions failed to take place. This is the way Newton viewed the solar system,[3] and, as we'll see, it's the way a number of proponents of an analogical version of the design argument see the universe.

Appeals to intuitions of *conspiracy* are as old as any of the others, at least in the written record. These intuitions arise from the observation that conditions in the world are just right for life as we know it to be possible – were things ever so slightly different then humans could not thrive. The Greeks, as we will see, noted that the environment of the Aegean brings cool winds just at the time of year when the sun threatens to scorch their crops.[4] Modern cosmologists point out that if certain physical constants had slightly different values, atoms would be impossible or the universe would have imploded long ago. The intuition in either case is that such facts are the consequence of a cosmic conspiracy – someone has arranged the world so that we can live in it. We will see this intuition developed into a detailed and very modern design argument in Chapters 15 and 16.

Below are four case studies intended to highlight each of the sorts of intuition discussed above. In the first case, we are quite certain that the object in question is the product of design. In the remaining three, this fact is precisely what the various design arguments are supposed to settle.

---

[3] See the "General Scholium" of Newton's *Principia* (1995, 439–43), or Hurlbutt 1965, 7–8, for a modern gloss.

[4] I am referring to the mention of the Etesian winds in II.131 of Cicero's *De natura deorum* (discussed in Chapter 3).

1.3 Case study 1: the Antikythera Mechanism

In 1900, a coterie of sponge fishermen made an extraordinary discovery.[5] They had been diving for sponges in waters off North Africa, and were sailing east towards their home on the Greek island of Syme when they were caught in a violent squall. They were forced to seek shelter in the lee of a small island called Antikythera, a rocky promontory some six and a half miles long by two miles wide in the channel between Crete and Kythera (the name of the island means "Against Kythera"). Anchoring over a shallow shelf, the fishermen safely rode out the storm. When the sea had calmed, they decided to take advantage of their novel surroundings, and began diving for sponges. This was a demanding job even after the invention of the 'standard suit', a canvas outfit topped with a brass helmet that allowed divers to walk on the sea floor and stay submerged longer than their lungs alone would allow.[6] One diver, Elias Stadiatis, found something at a depth of 140 feet, but it wasn't a sea-sponge. In the muck of the sea floor he discovered the 160-foot-long remains of an ancient shipwreck. Though obscured by millennia of ocean deposits, he could see that the sunken ship had disgorged a treasure of amphorae (ceramic storage vessels) and statues of bronze and marble. Elias returned to the surface with an outsized bronze arm and dreams of riches.

After the captain of the tiny two-cutter fleet verified the find, the fishermen completed their voyage home, and promptly set about partying for six months. When it was finally time to think about business again, the fishermen consulted the Greek government about their find – presenting the bronze arm as evidence – and in no time a salvage operation was sent back to Antikythera. The sponge-divers themselves conducted the operation under the guidance of an archeologist, and were compensated for whatever they brought to the surface. Over a grueling nine months of labor that saw the death of one diver and permanent disability of two more, a spectacular array of artifacts were recovered. These included many bronze

---

[5] Unless otherwise noted, my account of the discovery of the Antikythera Mechanism is derived from Price 1974.

[6] Because of an imperfect understanding of decompression sickness, the introduction of the diving suit called a 'skafandro' actually made the sponge-diver's trade more risky. Between 1886 and 1910 there were some 10,000 deaths and 20,000 cases of paralysis among sponge-divers in the Aegean (Warn 2000, 37).

statues, such as the *Antikythera Youth*, and marble statues that were manu-factured in pieces to be assembled at their destination. These statues were apparently copies of fourth- and fifth-century BCE works, while the wreck itself dates to between 80 and 50 BCE. The ship was a commercial vessel carrying cargo from Asia Minor to Rome when it seems it was sunk by a squall much like that which the sponge fishermen had fled.[7]

Among the objects brought back to the National Museum of Greece from the shipwreck were many formless lumps of what appeared to be either weathered marble or corroded bronze. The divers had collected these spurious artifacts in case they were pieces of something important. These unidentifiable fragments were placed into a cage in the museum, and consulted repeatedly as the statues were being reconstructed to see if they might fit in somewhere. At some point, one of these lumps of cor-roded copper cracked open to reveal gears and the remnants of Greek script (see Figure 1.1). It was immediately recognized as an important arti-fact, an obvious contrivance of sophisticated engineering minds. This was the Antikythera Mechanism.[8]

Most of us, when confronted with a bronze assemblage of gears, shafts, and inscribed plates – no matter how corroded – would immediately infer that it was the handicraft of some human designer. For most of us, this inference is made intuitively, without any conscious consideration. There is little doubt in this case that our intuitions – in particular the identifi-cation of the object as an important contrivance by the staff members of the National Museum – are correct. But what is it that motivates such a strong conclusion of design? What features of the object evoke our intui-tions? Can these features alone, upon careful reflection, justify our intui-tive conclusions of design? Since this is a book about design *arguments*, we are not interested in the psychological question of how each of us actually comes to conclude that an object like the Antikythera Mechanism was designed. After all, there are many irrational ways for individual people

---

[7] There is the unsubstantiated possibility that the ship was carrying the possessions of Cicero, whom we'll meet in Chapter 3. From 79 to 77 BCE, he resided at the School of Posidonius on Rhodes, and would have been sending his baggage home at about the time the ship in question sank off Antikythera (Price 1974, 9).

[8] For the most recent reconstruction of the remarkably complex functioning of the Antikythera Mechanism, see Freeth *et al.* 2006. The device was a sophisticated com-puter for forecasting a variety of astronomical phenomena.

ROBERT CAMP
2012

Figure 1.1   A fragment of the Antikythera Mechanism. Illustration by Robert Camp

to arrive at particular beliefs. These psychological mechanisms don't help us decide the truth of the proposition we have come to believe in. For that purpose, we need an explicit argument – we are only interested in how one might rationally justify a conclusion of design. Furthermore, we need to figure out whether any of the justifications we can provide are suitable for objects that may have a non-human provenance. The whole point of the design arguments considered in this text is to establish the existence of one or more gods, not people.

In the case of the Antikythera Mechanism, the relevant intuitions seem to be those of form. Of course, if we include writing in the set of properties we're considering, then it is obvious how an inference to human design can be made. But let's ignore the text for a moment since we want to determine whether other properties – properties that occur in controversial cases such as living things or the universe as a whole – might justify an inference of design. I suspect the staff of the National Museum would have been just as excited had the Mechanism lacked any sort of annotation. But why?

Some of the ways we might start to justify our conclusion of design on the basis of the properties related to form are, like the presence of writing, not general enough to apply to the controversial cases addressed by design arguments. We might, for instance, point out that the only known sources of gears are human engineers. The Antikythera

Mechanism contains many gears and other parts that are recognizable from other known human contrivances, and so we might be making a simple inductive inference. However, this sort of inference requires us to be familiar with many uncontroversial cases first. To apply the same inference to argue for the existence of God, we would have to have a stockpile of objects that we already know to have been designed by God. This approach won't get us far. But it seems there are some more general features we might point to in the Antikythera Mechanism that suggest a designer: the fact that many of the parts have a regular geometric shape; the fact that in combination the parts each contribute to a complicated 'function'. Even if we don't know what the function is, it seems clear that removing or altering a part – for example cutting a tooth off a gear – would ruin the delicate causal chain between the revolution of a shaft and the motion of the indicator dials. These are the sorts of features intuitively linked to design that were most famously emphasized by William Paley (see Chapter 8).

1.4 Case study 2: the bombardier beetle

To prime our intuitions about purpose, we need only look to the world of living things. The plants and animals of familiar experience possess a great number of traits – called 'adaptations' – that equip them to live a particular sort of life. For example, the extraordinary sense of smell possessed by wolves, the long legs of the cheetah, and the gnawing teeth of the rodent are all instances of traits that, at first blush, strike many as the sort of thing an engineer would have produced if tasked with providing for these animals. One particularly striking example comes from the insect world. It is the bombardier beetle (see Figure 1.2).[9]

The common name 'bombardier beetle' actually refers to more than 500 species of *Carabidae*, the family of ground beetles that includes the shiny green tiger beetles and the sun beetles, both common to northern climes.[10] What sets the bombardiers apart – and earns them their curious nickname – is their manner of defense. When provoked, the beetle

---

[9] The bombardier beetle is a favorite example of proponents of 'Intelligent Design', a topic considered in Chapters 12 and 13. See, e.g., Behe 1996, 31–36.

[10] Beetles are an enormously diverse group of animals, with around 350,000 known species.

Figure 1.2   A Bombardier Beetle. Illustration by Robert Camp

explosively discharges a boiling spray of foul-smelling caustic fluid from its hindquarters. It aims this discharge at its foe with great accuracy. Each shot is accompanied by an audible popping sound, like that of a tiny bomb, hence the name bombardier.[11]

To understand how the bombardier beetle produces these little explosions, we need to look into its anatomy (Figure 1.3). The spray exits the beetle from a nozzle at the tip of its abdomen (the backend of the beetle). This nozzle connects internally to a hardened, thick-walled reaction chamber shaped like a Y. At the upper branches of the Y are valves leading into two large reservoirs. These large sac-like reservoirs are each filled by a secretory gland connected to it by a long, thin, coiled tube. The reservoirs are surrounded by muscular tissue, rather like your stomach.

Most of the time, the reservoirs are filled with what is essentially rocket fuel – a mixture of hydroquinones (the fuel) and hydrogen peroxide (a powerful oxidizer). When the beetle wants to fire its weapon, it contracts the muscles around the reservoirs, forcing this fluid into the reaction chamber. Here, the fluid from the reservoirs mixes with two sets of enzymes, biological molecules that act as chemical catalysts. Peroxide and hydroquinone normally do not react at room temperature. However, in the presence of the enzymes, they do so quickly and explosively. The hydroquinone–peroxide mixture is quickly converted into a few kinds of benzoquinones (nasty, corrosive irritants), oxygen, and water. In the process, the entire solution warms up to about 100° C and pressure builds

---

[11] This etymology is recounted in the anonymous *Dialogues on Entomology, in Which the Forms and Habits of Insects Are Familiarly Explained* (1819, 123). The relevant passage is quoted in Eisner 2003, 41.

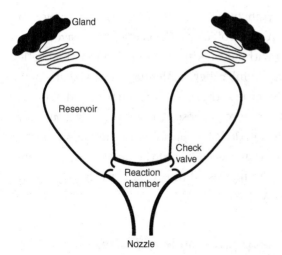

Figure 1.3   The parts of the Bombardier Beetle's explosive apparatus

within the reaction chamber. The valves leading back into the reservoirs are forced shut by the mounting pressure (they are the equivalent of what engineers call a 'check valve'), and the superheated solution is forced out the nozzle. As soon as a shot leaves the nozzle, the pressure drops and the valves from the reservoir reopen, admitting more fuel. This cycle repeats at a very high frequency until the beetle releases pressure on the reservoir. What sounds to a human like a single pop is really a rapid series of pulses.[12]

The mechanism of the bombardier's blast is remarkably similar to that of the pulse-jet, a sort of jet engine put to infamous use in Nazi Germany's V1 'flying bombs'.[13] In a pulse-jet, air is forced through a shutter and mixed with fuel. As it is compressed in the guts of the engine, the fuel–air mixture is ignited and the pressure shoots up. The shutters at the front of the engine slam shut, and the high-temperature, high-pressure mixture is forced out of the back. When the exhaust exits, the pressure drops and the shutters reopen. The cycle repeats as long as the engine moves forward and air is forced in. The result is a series of pulses that supply thrust.[14] This mechanism is in outline the same as the beetle's, only the beetle uses it to shoot ants.

[12] Chapter 1 of Eisner 2003 provides a thorough and thoroughly enjoyable overview of the anatomy and physiology of the bombardier beetle.
[13] Dean *et al.* 1990.   [14] See, e.g., Waltz 1950.

Aside from its striking resemblance to a piece of human technology – a resemblance which elicits intuitions of form – the beetle's defensive mechanism is difficult to discuss without making reference to *why* the beetle has these traits. The sense that bombardier beetles have organs that explosively eject caustic chemicals because they serve the purpose of defense is what I've been calling an intuition of purpose. An easy way to account for the curious equipment of the beetle is by appealing to the intention of a designer, an intelligent agent that has equipped the beetle with a defensive weapon. The question we're interested in is whether this easy explanation holds up under scrutiny.

1.5 Case study 3: gall wasps and their parasites

If you happen to live in a part of the world in which trees lose their leaves every winter, then it is almost impossible to avoid the handicraft of the gall wasps. If you carefully inspect the leaves upon the ground – particularly those of the oak tree if you live in North America – you are likely to discover numerous hard, woody bumps affixed to their desiccated veins (Figure 1.4). These wart-like growths are called 'galls'. Once you know what to look for, it's easy to find an enormous number of galls in a wide variety of shapes – some relatively formless, some smooth and spherical, some elaborately structured with spines or protrusions – affixed to plant leaves, stems, and twig tips. These galls are primarily produced by species belonging to two rather unrelated insect groups: the gall midges of the family *Cecidomyidae* (which are flies) and the gall wasps (of the family *Cynipidae*).[15] The galls you see on oak leaves are the work of gall wasps, and it is to the wasps that we will restrict our attention.

Each gall begins with the laying of an egg. In the case of a gall wasp, each female adult is equipped with a syringe-like appendage called an 'ovipositor' that protrudes from below her abdomen. With her ovipositor, she injects an egg into the tissue of a host plant, sometimes drilling through bark or woody stems. Once her larva hatches from its egg it begins to secrete 'morphogens', unknown compounds that reprogram the cells of the surrounding plant tissue to produce a home for the larva.[16] This is

---

[15] See Lanham 1964, 112–13, for a succinct overview of the gall-forming insects and the process of gall formation.

[16] Stone and Cook 1998.

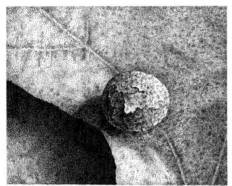

ROBERT CAMP
2012

Figure 1.4   A gall on the leaf of an oak tree. Illustration by Robert Camp

the gall. As the plant responds to the morphogens it forms an enclosure at the inside of which is a living chamber for the larva. This living chamber is lined with nutritive tissues on which the larva feeds. The larva has hijacked the developmental processes of the plant to the extent that it forces the plant to produce food for it throughout its entire juvenile development. Around this rich source of food, each species of cynipid wasp induces a different sort of protective structure. These outer structures may involve thick, hard wood, a complex series of air spaces, or coatings of sticky resin or fine hairs.[17]

The outer structure of a gall provides defense, primarily against other wasps. Just as the gall-making wasps parasitize their plant host, there are species of wasps that prey upon the gall-makers by penetrating the protective gall and laying their eggs in the body of the hapless larva within.[18] These gall wasp hunters have extra-long ovipositors that are used as drills. In some species, these ovipositors are much longer than the rest of the female's body. For instance, wasps in the genus *Megarhyssa*[19] have an ovipositor that is about 10 cm long, twice as long as the length of the female

[17] Stone and Cook 1998.
[18] In the *Eulophidae*, a family of wasps distinct from the cynipids, there are both species of gall-maker wasps and species that parasitize them (La Salle *et al.* 2009).
[19] The wasps of genus *Megarhyssa* actually parasitize horntail larvae – other structurally similar wasps in the same family parasitize the gall-makers. I use *Megarhyssa* as an example because the anatomy and drilling mechanism has been much more thoroughly examined for insects of this genus, in part because they are so large. What I say for *Megarhyssa* presumably holds also for the parasites of the gall-makers.

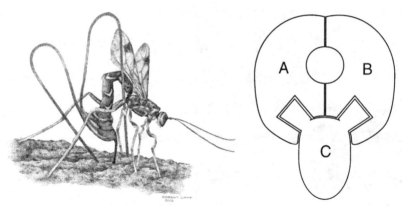

Figure 1.5   Sketch of a *Megarhyssa* wasp (left) inserting her long, needle-like ovipositor into a tree. Illustration by Robert Camp, based on a photo by Winnie Williams. On the right is a schematic cross-section of the thread-like ovipositor showing the two sliding filaments (A and B) and the fixed central filament (C) on which they move (see Abbott 1934)

from eyes to abdomen! In these wood-drilling parasitic wasps, the ovipositor is a hair-thin, flexible assembly of three filaments locked together and surrounding a membranous inner tube (see Figure 1.5). Two of these filaments slide along the third, which acts as a fixed rail. The sliding filaments grip the rail by way of a T-shaped groove – a striking instance of geometric simplicity and fodder for our intuitions of form. At the end of each of the two sliding filaments is a series of barbs.[20] To drill, the wasp pushes down on one sliding filament while pulling up with the other. Because the barbs are hard to pull back out of the wood, the side under tension locks in place and the side being pressed on drives deeper into the wood. Then the roles are reversed. In order to drill through thick, hard wood in this fashion without simply wearing the barbs off, these insects harden the drill tips by mixing metallic manganese with the chitin polymer of their cytoskeleton.[21] The result is an amazingly thin but durable and efficient drill.

Galls are the product of an extraordinary genetic manipulation. The drilling ovipositors of those wasps which parasitize their gall-making cousins are elegant drilling machines. With respect to both of these extraordinary adaptations, we are again struck by both the complexity of the biological equipment possessed by insects – after all, we haven't the slightest idea how to manipulate plant genetics like the tiny cynipid

20 See Abbott 1934; Riley 1888.   21 Quicke *et al.* 1998.

wasp does – and the apparent purpose for which these insects were so equipped. Why else would the larva possess the ability to excrete just the right compounds to have a tree build it a home unless it were so equipped with the *intention* that it do so? Why have a drill unless you were *intended* to drill through galls to make a living? The extraordinary, unpredictable, and ineffably complex ways in which organisms are adapted to survive and thrive continues to elicit intuitions of both form and purpose.

1.6 Case study 4: protein folds

Proteins are very large molecules, at least relative to most other molecules. They are ubiquitous components of living things. Proteins make up such things as hair, nails, and the mechanically resistant component of skin. At the cellular level, proteins are *the* molecules of life. A great variety of structures within cells are composed of proteins, such as the tiny 'cyto-skeleton' which maintains the cell's shape and the 'microtubule' railroad tracks along which the cell moves containers of chemicals. More impor-tantly, the myriad chemical reactions that take place within a cell are controlled and facilitated by a special class of proteins called 'enzymes'. Insofar as life is characterized by its complex chemical reactions, it is the class of proteins that makes it possible.

Roughly speaking, a protein is one or more chains linked together. Each chain in a protein is made up of twenty kinds of simpler molecule – the amino acids – that are strung together end to end like beads on a necklace. As a protein chain is built within the cell, it bunches or 'folds' into a stable, complex shape. These folded chains then assemble together as modules in a bigger structure – the protein. The mechanical and chemical properties of a protein are dictated precisely by this shape; there is a different shape for each distinct function. The proteins called 'enzymes' are especially dependent on shape. If you deform the relevant part of an enzyme just a little, then it can no longer facilitate or control the reaction it normally governs in the cell.[22]

It is something of a mystery how proteins fold. More accurately, it is extremely difficult to determine from the sequence of amino acids that make up each chain in a protein what shape that chain will adopt when

---

[22] For an overview of protein structure and enzyme functions, there are many excel-lent introductory biology texts. See, e.g., Curtis and Barnes 1989, 70–80.

it folds. Nonetheless, it seems to be the case that – whatever the details of folding – single protein chains only fold into a relatively small number of the shapes they could have if the only constraint were geometry.[23] From a strictly geometric standpoint, there are an enormous number of shapes any one amino acid chain might fold into. For instance, suppose we have a chain with only 150 links (this would be quite a small protein). To make things simple, imagine that each link can only snap into one of three positions relative to each of its neighbors. Even in this unrealistically restricted case, there would be some 10^{68} possible shapes for the folded chain.[24]

But here is where things get surprising and our intuitions of conspiracy begin to take hold. It seems that the laws of physics are just such that only 1,000 or so basic folded shapes are available for proteins, and these shapes are exactly those needed for the maintenance of life.[25] It seems as if the laws of physics are tuned to make sure that no matter which amino acids you string together, the resulting chain folds into one of a fixed number of shapes – an astronomically tiny proportion of all the shapes one can imagine for such a chain – and that these shapes are sufficient for life. Furthermore, the twenty amino acids out of which these folded chains are made are the most common in meteorites and the easiest to synthesize under prebiotic conditions – they seem to be available all over the place as the natural products of non-living processes. It is only a short step from these observations to the conclusion – or rather suspicion at this point – that the laws of physics were intentionally arranged such that the emergence of protein-based life would be possible or even overwhelmingly likely.[26] This sort of suspicion that the world is rigged for our benefit is what I've been calling an intuition of conspiracy.

1.7 Putting intuitions to the test

Intuitions focus our attention on certain possibilities, and in a pinch help us decide what to believe. But an intuitive or visceral sense that something is true of the world does not make it so. The preceding case studies were intended to prime your intuitions, to make the proposition that the world

---

[23] Denton *et al.* 2001.   [24] Denton *et al.* 2001, 332.   [25] Denton *et al.* 2001.

[26] In his book *Nature's Destiny* (1998), Michael Denton does go on to draw this further conclusion.

or various parts of it are the product of design appear plausible. But this is hardly sufficient to make the case that in fact organisms or the solar system or the laws of physics were so contrived. If it were, there would be no need of design *arguments* in the first place. Our task from this point on is to determine whether any of the facts which elicit intuitive conclusions of design – such as the adaptedness of living things or the improbable capacities of proteins – can function as premises in a sound argument for the existence of God.

2   Preliminaries

2.1 Overview

In Chapter 1, we considered some of the intuitions that suggest design in
the natural world. But intuitions alone are a thin justification for belief.
This book is concerned with the careful, rational justification that only
a good argument can provide. And history has provided plenty. Over the
past 2,000 years, a great variety of arguments have been offered, each of
which tries to establish that at least part of the world has been designed.
But before we can capitalize on these arguments – before we are in a
position to sift and weigh them, eliminating the weak and isolating the
strong – we have to do a little stage setting. We need to develop a vocabulary for referring unambiguously to the parts of an argument, and for
characterizing the different kinds of argument we come across. Not every
sort of argument – even when successful – offers the same degree of confidence in its conclusion, and we need to be able to recognize the difference.
Some of this work will be left to later chapters where fresh terminology
is introduced as it is needed. In this chapter, we'll consider a few simple
design arguments based on the case studies of Chapter 1 and introduce
just enough technical jargon to get us through our first set of historical
arguments.

2.2 Arguments and their parts

Consider the following argument for the existence of God based on features
of the parasitic wasps we considered in case study 3 of the last chapter:

Example 1
The egg-laying ovipositor of the wasp *Megarhyssa* is built just right to function
as an efficient drill. An efficient drill could not form by chance. Since every

structure is the result of design or chance, some intelligent being must have intended for the wasp to have such a drill. Therefore, there exists a being – call it God – who designed *Megarhyssa*.

The paragraph of Example 1 expresses something more than an intuition. I said before that an intuition is like a 'gut reaction'. These can generally be expressed as a single sentence like, "It sure seems as if bombardier beetles were given biological guns in order to shoot ants." Already you can see that an argument like the one above is something more complex. To begin with, it involves more than one claim. Furthermore, these claims are connected – they are intended to work together to support or justify the assertion that God exists. Some of the claims are offered as reasons to believe in the truth of another. This relation of support is what distinguishes an argument from a collection of mere assertions.

To be as unambiguous as possible, we can say that an argument is a collection of statements. *Statements* – also known as 'propositions' – are claims that are either true or false. They are typically expressible in a single sentence. For example, "Humans are mortal," "The orbit of Pluto is beyond that of Jupiter," and "Granite is denser than lead" are all statements, two of which are true and one of which is false. In Example 1, the sentence "The egg-laying ovipositor of the wasp *Megarhyssa* is built just right to function as an efficient drill," is a statement. In every argument, one of the statements is special – the conclusion. The *conclusion* is that statement the truth of which a given argument seeks to establish. It is the claim in question, the claim that motivates giving an argument in the first place. If we weren't interested in the truth of the conclusion, we wouldn't bother with an argument at all. In most design arguments, including Example 1, the conclusion is the claim "God exists."

The rest of the statements in an argument are there to support the conclusion. As I said, an argument for a conclusion is different from a mere assertion of the conclusion because the argument offers reasons to believe the conclusion is true. Of course, each of the claims we offer in support of a conclusion might itself require some support. For instance, it may not be obvious that an efficient drill could not develop through some chance process, and one might have to support this claim with further statements if the argument is to be successful. But we can't go on giving reasons indefinitely. To support any claim, we ultimately need some statements to serve as a foundation – we need premises. *Premises* are those statements in an argument that are offered as true without providing additional support. So for instance, in Example 1, the statement "An efficient drill could not form by chance," is asserted without providing any explicit reasons to think this is so. That doesn't mean that no such reasons could be offered – it just means that none are given. In the shortest arguments, there are only one or a few premises, and these directly support the conclusion without any other statements. In more complex arguments, the premises support what are called *sub-conclusions*, which in turn support the conclusion of the argument. In this way, arguments can be assembled from other arguments. In Example 1, for instance, there is a single sub-conclusion: "Some intelligent being must have intended for the wasp to have such a drill." This statement is the conclusion of a miniature argument made up of the statements in the first three sentences. It serves in turn as support for the conclusion that "there exists a being – call it God – who designed *Megarhyssa*."

2.3 Inferences, validity, and soundness

All I've said so far is that arguments contain statements offered in support of a conclusion. But we haven't said anything about what makes an argument convincing. Since the whole point of our inquiry is to assess whether any design arguments are convincing, we need a rigorous way to approach such an assessment. This means looking closer at the relation of support. Unfortunately, it also means introducing a bit more jargon, beginning with the term 'inference'. An *inference* is the affirmation of one statement on the basis of others. For instance, asserting that "Socrates is mortal" because "All men are mortal" and "Socrates is a man" is an example of an inference. Inferences link together the statements in an argument, connecting premises to conclusions or sub-conclusions. There are different kinds of inferences – different patterns according to which one statement is affirmed on the basis of others – and these different kinds give us varying degrees of confidence in the statement being affirmed. The strongest sort of inference in this respect is called *deductive*. For a deductive inference, the truth of one statement is supposed to be guaranteed by the truth of those offered in its support. An argument whose statements are connected entirely by deductive inferences is said to be deductive, and the truth of the conclusion is intended to follow necessarily from the truth of the premises – there is no room for doubt. Consider, for instance, the following argument:

Example 2
Premise 1: If something was designed, then it must have a designer.
Premise 2: The bombardier beetle was designed.
Conclusion: The bombardier beetle has a designer.

This is an example of a deductive inference. It is such a common sort of deductive inference that it has its own name, *modus ponens*, which is Latin for "mood that affirms." Instances of modus ponens are always of the form 'if p then q; p; therefore q', and they are deductive. If Premise 1 and Premise 2 are true, it is supposed to follow necessarily that the Conclusion is true.

Just because an argument is deductive, however, does not make it good – it just means that the truth of one statement is *supposed* to follow with certainty from the truth of the others. In the case of modus ponens, this intended relation between premises and conclusion really does hold. If the two premises are true, then necessarily the conclusion is true. All successful deductive inferences with this property – inferences for which the truth of the premises really does necessitate the truth of the conclusion – are called *valid*. Defective deductive inferences – inferences intended to be deductive but for which the premises fail to support the conclusion – are called *invalid*. Validity is necessary for a deductive argument to be convincing but it is still not enough to guarantee the truth of its conclusion. What makes an argument valid is a relation between the premises and the conclusion: if the premises are true, then necessarily the conclusion is too. But this relation can hold even if the premises are false – an argument can be valid yet rest on false premises. The argument above, for instance, is valid, and it remains so even if it happens to be false that the bombardier beetle in question was designed. A *sound* argument is a valid argument with true premises. If a deductive argument is sound, then the conclusion must be true. The argument of Example 2 would be sound if in fact we knew the beetle to have been designed and if in fact Premise 1 is true (which it surely is merely by dint of the definition of 'designed'). In that case, we would know with certainty that beetles have designers.

There are areas of inquiry where substantial deductive arguments abound. Mathematical proofs, for instance, are all deductive arguments, and a correct proof is a sound deductive argument. A mathematical proof confers certainty in the truth of the theorem it supports. For example, it is necessarily true that the sum of the interior angles of a plane triangle is 180°, since we can construct a valid deductive argument (a geometric proof) for this fact from true premises about plane geometry. Outside of mathematics, deductive inference is the gold standard for argumentation. We'd like to know everything with the same certainty we have about geometric theorems, so we support conclusions with deductive arguments whenever possible. However, sound deductive arguments are often unavailable, and so we must look beyond deductive inference in order to support our claims. In fact, the majority of design arguments we'll meet are not deductive.

2.4 Inductive arguments

Outside of deductive inference there is a zoo of more poorly defined inference forms which can be lumped together as *inductive*. The term 'induction' has both a broad and narrow sense.[1] In the narrow sense, it refers to the support of a general proposition by enumeration of instances in which the proposition holds. So, for example, I might support the claim that all ravens are black by pointing to very many instances of black ravens. In the broad sense in which I use the term, induction refers to any inference in which the conclusion cannot be deduced with certainty from the premises – in some sense, there is more in the conclusion than was contained in the premises. In this broad sense, inductive inferences can be put in the general form '*p* therefore *q*', but the truth of *p* can only confer a degree of confidence less than complete certainty on the truth of the conclusion *q*. Here's an example drawn from case study 4:

Example 3
Premise: All the proteins we've studied have one of a relatively small number of shapes.

Conclusion: All proteins have one of these shapes.

---

[1] See, e.g., Audi 1999.

And here's another based on case study 1:

Example 4
Premise: The lump of bronze recovered from Antikythera was found near a shipwreck.
Conclusion: The lump of bronze was designed by humans.

In both of these examples, even if the premise p is true it is possible for the conclusion q to be false. In the case of the protein shapes, it might be that we happened to have studied all of the proteins with the special shapes first and that the remaining proteins yet unchecked will have very different shapes. It's not likely, but it is possible. Similarly, the presence of a shipwreck nearby doesn't guarantee that a lump of copper was crafted by people. It might have been present on the ocean floor for altogether different reasons before the ship even sank. Nonetheless, if the premise in each argument is true it is rational to increase our confidence in the truth of the conclusion. If I notice a large shipwreck full of curious objects near a strange metallic lump on the sea floor, I will think it more likely that the lump was the result of human manufacture than if I hadn't seen the shipwreck. It is in this very general sense that I mean an argument is inductive. All inductive arguments involve extrapolating from some fact about the world to places and times not in evidence. Some proteins have a special shape, so probably they all do. The copper lump and the shipwreck are associated now, so it is likely that they were associated earlier. It is important to emphasize that inductive arguments cannot yield certainty. Unlike a mathematical proof, the best we can get out of an inductive argument is a great deal of confidence in its conclusion.

Another important feature to note about inductive inferences is that premises can independently add cumulative support to a conclusion. This is not the case for deductive arguments. If the conclusion necessarily follows from the premises in a deductive fashion, then there is nothing gained by adding more premises – the conclusion is no more or less certain. If, however, we add premises to an inductive argument where each new premise independently supports the conclusion, then the conclusion can be made more certain or probable. For instance, if I tell you that I cracked open the copper lump and discovered symmetric structures like wheels, and that the copper was found to be unusually pure, then the conclusion of human design becomes much more probable than it would be given any of those facts in isolation.

The notions of validity and soundness do not apply to inductive arguments, since the conclusion of such an argument is not supposed to follow with certainty.[2] Rather, we can speak of the analogous properties of strength and cogency. A *strong* inductive argument has the following property: if its premises are true, then the conclusion is highly probable or credible. This is analogous to validity. If the premises of a strong inductive argument are actually true, then we say the argument is *cogent*. A cogent argument makes the truth of its conclusion highly credible or probable. Cogency is analogous to soundness.

2.5 Argument diagramming

Arguments are tools for supporting claims. But how do we determine whether a particular argument successfully supports a claim, and to what degree? In order to decide this question and apportion confidence in the conclusion, we need to assess two aspects of an argument: (i) the truth of its premises, and (ii) the validity or strength of its inferences. If the premises are true and the inferences are valid or strong, then we can be more or less certain of the truth of the conclusion. In order to consider these aspects of a complex argument, it is often helpful to represent the argument as a diagram. An *argument diagram* is a two-dimensional visual representation of an argument. In a diagram, every statement of the argument is represented with a box, and every inference with an arrow.[3] Consider, for example, the argument of Example 2. Figure 2.1 presents this same argument as a diagram. Each box of the diagram contains a statement – a claim that is either true or false. Boxes are connected by arrows which represent inferences. The box to which an arrow points is a conclusion (or sub-conclusion if it gets used in turn as support for another inference). In Figure 2.1, the box labeled (C) contains the conclusion of the argument. The boxes from which arrows originate are supposed to lend support for the truth of the conclusion via an inference. Note that in this

[2] Some authors treat inductive arguments as invalid; I prefer to emphasize the fact that inductive arguments have an altogether different nature, and so abstain from using the terminology appropriate for deductive arguments.

[3] I follow the diagramming syntax used in Harrell 2010.

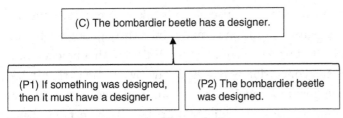

Figure 2.1   The argument of Example 2

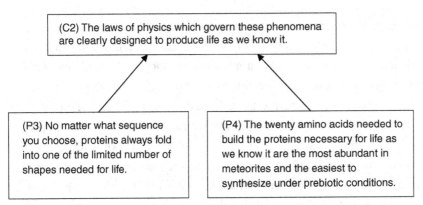

Figure 2.2   The argument of Example 5

case, two statements – the premises (P1) and (P2) – are bound together with a bracket and connected to the conclusion with a single arrow. This indicates that the combination of the two premises supports the conclusion, regardless of whether each premise does so on its own. In this example, the inference indicated by the arrow is deductive, and the conclusion follows with certainty from the linked premises.

For comparison, consider the following argument:

Example 5
No matter what sequence you choose, proteins always fold into one of the limited number of shapes needed for life. Furthermore, the twenty amino acids needed to build the proteins necessary for life as we know it are the most abundant in meteorites and the easiest to synthesize under prebiotic conditions. Thus, the laws of physics which govern these phenomena are clearly designed to produce life as we know it.

This argument is diagramed in Figure 2.2. In this case, there is an arrow from each of the premises (P3) and (P4) to the conclusion (C2). This represents the fact that each premise alone is taken to provide support for the conclusion. This convention is maintained throughout the book: premises that jointly support a statement are linked with a bracket and connected to the sub-conclusion with a single arrow; premises that independently lend support to a conclusion as in an inductive argument are linked to the sub-conclusion by individual arrows. Both sorts of arguments can be combined into more complex argument structures, as we'll see.

Ultimately, diagramming arguments makes them easier to assess. A complete argument diagram lays out all the elements of an argument for straightforward examination: premises, sub-conclusions, and the conclusion are given as boxes, and inferences as arrows. As I said above, one thing we want to do when assessing an argument is to make sure the premises are true. From our argument diagram, we can quickly identify the premises – the claims on which an argument rests – by looking for the boxes with no arrows leading into them. We can then consider each of these in turn and determine whether they are self-evidently true, empirically verifiable, or supportable by an additional argument. The other aspect of argument analysis involves an examination of the inferences. These again are transparently represented by arrows in the diagram, so the task reduces to asking whether the statements connected to a conclusion by an arrow really would support that conclusion if they were true. In short, constructing an argument diagram makes the skeleton of an argument explicit. This in turn makes critiquing an argument easier. We will put this technique to use in the next chapter when we begin our historical survey of design arguments.

2.6 What is a design argument?

Before we can make a survey of design arguments, we have to know what we're looking for. I've said quite a lot about arguments and how to represent them, but I haven't said anything yet about what makes an argument a *design* argument. To do so, I need to introduce one more distinction: some arguments are 'a posteriori' and some are 'a priori'. An argument is a posteriori (a Latin phrase meaning "from what comes after") if one or more of its premises appeal to a fact of experience. So, for example, one of the arguments above appeals to the small number of shapes into which proteins are seen to fold. Since it uses a premise that appeals to experience, it is an a posteriori argument. So too is any argument that appeals to the combined mass of humans presently on Earth, or to the shape of the state of Pennsylvania, or to any other fact that might have been otherwise. Any argument that takes as a premise a fact about the world as it happens to be – a fact that can only be learned through experience – is a posteriori.

An argument that is a priori ("from what is before") is one containing no such appeals to experience. All of the premises in an a priori argument are taken to be true necessarily or as a matter of convention. They are supposed to be true independent of the way our particular world has turned out. Mathematical proofs are examples of a priori arguments since, for instance, the fact that there are no prime numbers between seven and eleven can be justified without any appeal to experience.[4]

Design arguments are a posteriori arguments that appeal to facts of our experience which suggest the intentional contrivance of an intelligent agent. There are a great many experiential facts that have been associated with design, and much of the work in making a design argument involves supporting the connection between some observable state of affairs and the action of an intelligent agent. Once a connection between special properties of things in the world and intelligent agency has been established, one can then infer the existence of one or more designers – usually

---

[4] This is the modern way of viewing the distinction. In the philosophy of Aristotle and the scholastics, an a priori argument was one in which the conclusion was justified by appeal to the existence of the cause or ground of the state of affairs it described (Lacey 1996, 14). That is, were I to argue that I have extension because I am a material being, this would be a priori. A posteriori arguments were those which argued for the truth of a proposition on the basis of its effects or consequences. Inferring a designer from marks of design is such an inference. In Question 2, Article 2, Part 1 of the *Summa theologica*, Thomas Aquinas put it this way: "Demonstration can be made in two ways: one is through the cause, and is called *a priori*, and this is to argue from what is prior absolutely. The other is through the effect, and is called a demonstration *a posteriori*, this is to argue from what is prior relatively only to us. When an effect is better known to us than its cause, from the effect we proceed to the knowledge of the cause. And from every effect the existence of its proper cause can be demonstrated, so long as its effects are better known to us; because since every effect depends upon its cause, if the effect exists, the cause must pre-exist" (Aquinas 1920, 12).

3 Arguments from antiquity

3.1 The emergence of design arguments

We'll begin our historical survey of design arguments in the first century BCE. This choice is somewhat arbitrary, since by that point in history, what we would recognize as a design argument had been around for centuries. But it was around this time that Cicero wrote his *De natura deorum*, the oldest surviving manuscript to gather together and assess the full crop of ancient design arguments. This work represents the earliest point at which the handful of arguments we will follow over the next two millennia were indisputably in play, and so it makes a natural starting point for studying design arguments. Nonetheless, it is worth glancing back into the earlier days of design arguments – and of philosophy itself – before we go forward with our survey.

It would take another book altogether to trace the crystallization of recognizable design arguments from the rich brine of early philosophy.[1] To begin with, the emergence of design arguments had to wait for a particular combination of philosophical views to develop.[2] This was not for lack of conviction in the existence of gods or divine intelligence. Rather, the earliest known attempts at systematic, scientific explanation of the world (as opposed to mythic or religious ones) simply assumed that intelligence like ours is the dominant mechanism in the physical world – everything was considered animate to one degree or another. Particularly for the scientifically minded Milesians, early attempts at a workable physics focused on the search for 'principles', the primary causes of all structure

---

[1] Fortunately, one such book exists: Sedley 2007. For a broader treatment of ancient Greek natural theology, see Gerson 1994.

[2] The broad account given here of the conditions necessary for design arguments to emerge parallels Sedley 2007.

and change. For instance, Thales of Miletus declared that the principle is water. By this he seems to have meant that water is the original substance of the world, and all subsequent physical features of the world can be accounted for on the basis of the properties of water and its changing states. If we are to believe Aristotle, Thales also said that "all things are full of a gods."[3] That seems an odd thing to say after offering a sort of physical theory of the world, but it is characteristic of early accounts. It is a bit of a caricature, but such accounts understood living intelligence to be a pervasive force which could be used to explain the structure of the world and our experience of it. Within such a view there is no need of a design argument – if you already believe that the only causes of change in the world are animate (and thus to some extent intelligent) then it would be a waste of your time to argue for the existence of divine intelligence. In order for such an argument to be necessary, someone had to first assert a distinction between mind and matter, between animate and inanimate causes. Someone had to raise the serious possibility that other causes of structure and change are operative in the world besides those we identify with life and mind.

Anaxagoras has been credited as the first to offer a physics that treats mind as the craftsman of independent matter.[4] His cosmogony (a cosmogony is an account of the origin of the world) invokes both a cosmic intelligence called *nous* and a quantity of inanimate materials composed of the elements earth, air, fire, and water as well as the 'seeds' of biological life. Importantly, these raw materials on which *nous* operates are treated as having definite causal properties of their own – *nous* operates by planning outcomes and manipulating matter to achieve them. In this sense, Anaxagoras gives us the first clearly teleological account – matter is governed by its own principles but is shaped and formed to meet the goals of an independent intelligence. It is also what is known as a 'dualist' account, meaning an account that posits two basic kinds of stuff. Anaxagoras' theory is dualist in the sense that it distinguishes physical and mental causes.

---

[3] *De anima*, 411a, 8–9. See, e.g., McKeon 1941, 553.

[4] See Sedley 2007, ch. 1. See also Dryden and Clough 1876, 108, where Plutarch claims that Anaxagoras "was the first of the philosophers who did not refer the first ordering of the world to fortune or chance, nor to necessity or compulsion, but to a pure, unadulterated intelligence, which in all other existing mixed and compound things acts as a principle of discrimination, and of combination of like with like."

The cosmogony of Anaxagoras was followed by more elaborate dualist accounts of the world and its structure, such as that of Empedocles.

By the end of the pre-Socratic era, these rather successful dualist accounts were being challenged by the materialist theories of the atomists. For the atomists, mind is not independent of matter but is, like everything else, a consequence of the properties and motions of material atoms. Atomism began with Leucippus, though his more prolific student Democritus is better known. The atomists posited that the world is made of atoms and void. The atoms – indivisible, partless bits of matter that come in every shape and size – make up every material object in the world. Change is accounted for by the changing configuration of these atoms as they move through the void. The atomists asserted that the universe is infinite in extent and contains infinitely many randomly moving atoms. By appealing to infinite time and an infinite supply of atoms, the atomists could account for the structure of the world purely in terms of what we would today call 'statistical necessity'.

So by the mid fifth century BCE the stage was set for the appearance of design arguments. Two conflicting views were in place: in one view, the world is shaped by a divine intelligence; in the other, there are only atoms. In order to favor the first, one needed an argument against the materialists – one needed a design argument. It seems that, perhaps not surprisingly, Socrates gave us the first such argument.[5] Since we'll see this argument reflected in the Stoic arguments below, I won't analyze it in detail here. But in outline, Socrates says that whatever exists for beneficial purposes must be the result of reason, not of chance. He then points to a number of biological adaptations such as the eye that, by his reckoning, clearly serve beneficial purposes such as seeing. From this, he concludes that a god exists. He then goes on to describe the numerous ways in which the world has been crafted for the apparent benefit of mankind, particularly for human pleasure. He uses these facts to argue that not only was the world created, but the creator has special concern for man.

The design argument of Socrates – recounted only in the works of Xenophon – is not particularly well known. Much more famous is the

---

[5] The argument recounted here appears in Xenophon's *Memorabilia* (Book I, ch. iv) (Kühner 1847, 32–38). See also Sedley 2007, 84–85. Some credit Diogenes of Appolonia with the first design argument (see, e.g., Pease 1941). I follow Sedley 2007 in giving that distinction to Socrates.

argument of Socrates' protégé, Plato, though it is of a very different character. At the outset of his dialogue *Timaeus*, Plato offers a design argument that, like the deductive argument of Thomas Aquinas we studied in Chapter 2, appeals only to the most general features of the world. In particular, Plato argues (via the character Timaeus) that the world is sensible and so must be subject to change and must at some point have come to be. Anything that 'becomes' (as opposed to things which exist unchanging forever) must have a cause, and that cause must be the craftsman of the universe. Again, I do not wish to look into Plato's argument in detail, only to point out that he gives us a mature design argument – an argument from features of the world to the conclusion that there exists an organizing, designing craftsman of the world.

By the mid first century BCE the debate had taken the form it would retain – at least in its broad strokes – until the present day. By this time, there were three principal philosophical schools, each with roots reaching back centuries to the zenith of Greek philosophy. There were the Stoics, who perpetuated something like Anaxagoras' view of a material world shaped by an intelligent and omnipresent world-soul. They presented a suite of design arguments in favor of the existence of a divine intelligence, capable of causing change independent of the properties of matter. The Epicurean atomists on the other hand were strict materialists. They viewed the world – gods included – as entirely composed of atoms and void. They argued that what the Stoics perceived as evidence of design can be explained by chance and an infinite supply of moving atoms. While the Epicureans admitted the existence of gods, these gods were supposed to be entirely set apart from the world of human affairs, having nothing to do with people and bearing no responsibility for the world's construction. For our purposes, the Epicureans may as well be viewed as defending the atheist position. Finally, there were the Academics, who claimed descent from Plato's Academy. At that stage in their history, the Academics represented skepticism, denying that we can have any certain knowledge of the gods, of atoms, or of the absence of either. We are most interested in the Stoic position, and in section 3.4 we will consider the Stoic design arguments as the Roman Cicero presents them. But before we get to the arguments we should consider our source: just who was Cicero and why did he record these early design arguments?

3.2 Cicero and the Hellenistic world

Marcus Tullius Cicero – or 'Tully' as he is affectionately known to many – lived from January 3, 106 BCE to December 7, 43 BCE.[6] Forty years before Cicero was born, the Greek empire of Macedon was defeated for the last time and all of Greece became a province of the Roman empire.[7] The world of independent Greek city-states which had given rise to the flowering of Hellenic speculative thought had long since crumbled under the conquering boot of Alexander the Great, and by the time of Cicero's birth the Hellenic world had lost its political and cultural independence altogether.[8] The intellectual influence of the Greeks, however, remained strong, though the focus of their Latin conquerors was markedly different. The Romans readily assimilated the various philosophical schools of the Greeks, though they had relatively little taste for metaphysics or cosmology. Rather, the Romans saw it as "the philosopher's task to provide the individual with a code of conduct which would enable him to pilot his way through the sea of life, maintaining a consistency of principle and action based on a certain spiritual and moral independence. Hence the phenomenon of philosopher-directors, who performed a task somewhat analogous to that of the spiritual director as known to the Christian world."[9]

According to Cicero, it was for such practical concerns over moral conduct that he wrote *De natura deorum* (The nature of the gods):[10]

> We must perpetuate religious belief, which is closely associated with the knowledge of nature, but we must likewise eradicate superstition root and branch. Superstition breathes down your neck, following you at every turn. You may be listening to a seer or an omen; you may be offering a sacrifice or observing a bird's flight; you may be confronting an astrologer or soothsayer. It may be lightning or thunder or a bolt from heaven; it may be the birth or creation of some prodigy. Something of this sort must necessarily occur almost every day, so that our minds can never be at rest.

Cicero thought that his treatise on the nature of divinity would go some way to clearing away superstition from proper religious sentiment and

---

[6] Copleston 1993, Vol. I, 418.   [7] Langer 1948, 80.
[8] See Copleston 1993, Vol. I, 379–84.   [9] Copleston 1993, Vol. I, 380.
[10] From Cicero's *On Divination* and quoted by P. G. Walsh in the introduction to Cicero 1998, pp. xxvi–xxvii.

practice. He faced the problem of embracing the traditional state religion of the Roman republic while simultaneously conceding that it provides fertile ground for pernicious superstition. His interest in the philosophical treatment of the existence of the gods appears then to have been one of practical concern, not theoretical curiosity. If one can provide an account of what the gods are and why we should believe them to exist, then false beliefs about, say, fortune-telling would be easy to distinguish and discard.

Cicero was mostly famous for his skill as an orator in the courts, not as a philosopher. This remains true in our time as well. He devoted a significant portion of his life to public office, though he never attained a rank commensurate with his ambition – he was, among other things, a consul and governor of the eastern Roman province of Cilicia, but never a senator.[11] His philosophical education was eclectic in the modern sense. He studied with representatives of each of the major philosophical traditions, including the Epicureans Phaedrus and Zeno of Sidon, the Stoics Diodotus and Posidonius, and the successive heads of the Academic school, Philo and Antiochus.[12] So by the time he was 30, Cicero had acquired a comprehensive education in Greek philosophy as it existed under Roman rule.[13] But apparently he was not entirely convinced by any one of his illustrious teachers. The philosophy Cicero adopted was a hodgepodge of doctrines from the several schools, strung together with little or no concern for consistency. This was characteristic of the Roman philosophical school known as the 'Eclectics'.[14]

While he continued to study philosophy throughout his life, Cicero produced most of his philosophical writings in his last few years, a period in which he produced a remarkable number of books.[15] It is perhaps not surprising, given his philosophical inclinations, that these works were largely derivative,[16] made up mostly of surveys of the thought of the various philosophical schools on different topics. The skill with which Cicero made these abstruse doctrines comprehensible to his Latin audience, however, is laudable. It is also a boon to us, since many of the texts Cicero quotes, cites, or summarizes have since been lost to the modern world – his accounts are amongst the few we have available to reconstruct the contents of such

[11] Cicero 1998.   [12] Cicero 1998, 418; Copleston 1993, Vol. I.
[13] Cicero 1998, xiv.   [14] Delfgaauw 1968, 39–40.
[15] Copleston 1993, Vol. I, 418–19.   [16] Copleston 1993, Vol. I; Cicero 1998.

works as Aristotle's *De philosophia*.[17] Of greatest relevance to us is the fact that Cicero's *De natura deorum* provides a thorough statement of the theological stance of the Stoics.

3.3 The Stoic school

So who were the 'Stoics'? The school was founded by Zeno[18] around 300 BCE. He lectured in the Στοά Ποικίλη – or 'Stoa Poikilê' when anglicized – which means 'colored portico' in classical Greek. It was from this word 'stoa' that the school obtained its name.[19] Zeno was succeeded as head of the school by Cleanthes – a name we'll see again – who in turn was succeeded by Chrysippus.[20]

Though the Stoics saw the conduct of life – and thus the study of ethics – as the primary concern of philosophy, their epistemological and metaphysical doctrines are more important for our purposes. With respect to epistemology, the Stoics believed that all knowledge derives from sense experience. More specifically, they thought that particular objects make an impression on the soul which remains behind as a memory after the immediate act of perception. Knowledge concerns these impressions and their recollections.[21] Given this emphasis on perception, it is not surprising, then, that the Stoic arguments for the existence of god are – as we'll see below – a posteriori, appealing to perceptible features of the world.

While the Stoics were empiricists in the sense that they believed that we are born with blank souls that are later filled with sense impressions, they also thought that "it is only through Reason that the system of Reality can be known."[22] They seemed to think we possess certain innate concepts. It is partially on the basis of such concepts that the cosmology at the core of Stoic doctrine was constructed. I'll have more to say about this

---

[17] See, e.g., nn. 42 and 44 on pp. 177 and 178 of Cicero 1998, respectively.

[18] This is not the same Zeno whose paradoxes, including Achilles and the Tortoise, you may know. That Zeno was from Elea and was a student of Parmenides. He was born around 489 BCE (Copleston 1993, Vol. I, 54).

[19] The Greek phrase is from Copleston 1993, Vol. I, 385, but the translation is my own. Baltzly says that "[t]he name derives from the porch (*stoa poikilê*) in the Agora at Athens decorated with mural paintings, where the members of the school congregated, and their lectures were held" (Baltzly 2008).

[20] Copleston 1993, Vol. I, 385.   [21] Copleston 1993, Vol. I, 386–87.   [22] Copleston 1993, Vol. I, 387.

cosmology – the Stoic theory of the origin, structure, and nature of the world – after we look at their various design arguments. So let us turn our attention to the latter, and greet the arguments whose descendants continue to spur debate today.

3.4 The Stoic design arguments

The Stoic arguments from design appear in Book II of *De natura deorum*.[23] Specifically, versions of the argument appear in II.4–5, 43, 49–56, 89–90, 93–94, and 120–54.[24] Supporting premises and reformulations of each version are scattered throughout Book II. Cicero's account of the Stoic arguments are delivered through the character of Quintus Lucilius Balbus.[25] We can distill Balbus' lengthy discourse into four principal design arguments: (i) the argument from order, (ii) the argument from purpose, (iii) the argument from providence, and (iv) the argument by analogy.[26] The *argument from order* is first stated, albeit briefly, in II.4. It goes something like this: the heavenly bodies exhibit order in their motion, therefore the motion of the heavenly bodies is the product of intelligent agency. In II.49–56 this argument is elaborated, with various celestial motions described in detail as examples of the sort of intricate regularity exhibited by the cosmos. Here again, the argument is left largely implicit.

Perhaps the most explicit presentation is provided in II.89–91 by way of a parable. Balbus recalls a character from a play by Accius. The character is a

---

[23] For a different take on the arguments of *De natura*, Book II, see Gerson 1994, ch. 4.2.

[24] These numbers refer to passages in the text. Ancient texts have been adorned by modern scholars with such numbers so that, irrespective of whether one is reading the text in its original language or a translation, we can all refer unambiguously to the same passage. Cicero's *De natura deorum* is divided into three books (labeled with roman numerals), and each of these in turn is divided into numbered passages. So II.17 refers to Book II, passage 17. Sometimes I provide page numbers in a footnote where particular quotations can be found, but these page numbers refer only to the specific translation I have used, namely Cicero 1998.

[25] The character Cotta is Cicero's mouthpiece in the dialogues. Cotta represents the skeptical Academic school with which Cicero identifies himself.

[26] Some authors consider a fifth type of argument, taking the fact that the world contains beauty as a premise. Indeed, Balbus says in II.98 that "At this point we can abandon the refinements of argument, and concentrate our gaze, so to say, on the beauty of things which we declare have been established by divine providence." But I would argue that this is merely an extension of the argument from providence, and so will group the two (beauty and providence) together.

shepherd who, never having seen a ship before, is astonished and confused when he spots the ship of the Argonauts from a distant mountaintop. At first, he wonders what sort of thing this ship may be. He assumes it is some lifeless freak of weather or ocean processes. But as it draws closer and the shepherd hears tunes being sung and, eventually, sees men walking about on the deck, he realizes that it is a vessel driven by intelligent beings. Balbus tells us the lesson we are to draw from the parable and from the preceding arguments:[27]

> If the first sight of the universe happened to throw them into confusion, once they observed its measured, steady movements, and noted that all its parts were governed by established order and unchangeable regularity, they ought to have realized that in this divine dwelling in the heavens was one who was not merely a resident but also a ruler, controller, and so to say the architect of this great structural project.

This argument is made somewhat more explicit a little further on:[28]

> When we observe that some object – an orrery, say, or a clock, or lots of other such things – is moved by some mechanism, we have no doubt that reason lies behind such devices; so when we note the thrust and remarkable speed with which the heavens revolve, and preserving the whole of creation in perfect safety, do we hesitate to acknowledge that this is achieved not merely by reason, but by reason which is preeminent and divine?

With this final reformulation, we have almost enough to reconstruct a valid argument. What's missing in the text, however, is the key premise that order is always a consequence of intelligent agency. If we grant this additional premise as implicit, then we can reconstruct a valid argument from order as shown in Figure 3.1. (In this figure, the argument is represented as the sort of diagram introduced in Chapter 2.)

Of course, we don't yet have reason to believe that order is always the result of intelligent agency and so we do not yet know whether the argument is sound. In another passage of the dialogue, Balbus provides an important clue as to how we might support this vital premise. In II.93 he expresses disbelief that anyone could think that chance collisions of atoms created a universe of great complexity and beauty. He says that:

> [A]nyone who imagines that this could have happened, must logically believe that if countless numbers of the twenty-one letters of the

---

[27] II.90, see Cicero 1998, 79.   [28] II.97, see Cicero 1998, 92.

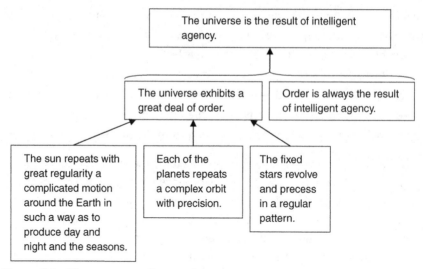

Figure 3.1   The argument from order

alphabet, fashioned in gold or in some other substance, were thrown into the same receptacle and then shaken out upon the ground, they could form the *Annals* of Ennius made immediately readable before our eyes. Yet I doubt if as much as one line could be so assembled by chance.

We can replace this with a more modern example, one due to Fred Hoyle:[29]

> A junkyard contains all the bits and pieces of a Boeing 747, dismembered and in disarray. A whirlwind happens to blow through the yard. What is the chance that after its passage a fully assembled 747, ready to fly, will be found standing there? So small as to be negligible, even if a tornado were to blow through enough junkyards to fill the whole Universe.

The claim in both cases is that chance simply cannot produce certain sorts of order, and this is something we know from experience. If this is true, it is still not quite enough to justify the claim that order comes only from intelligence. To get there we need to invoke an idea about causes that is implicit in the debate between the atomist Epicureans and the dualist Stoics, namely that every event is either the outcome of chance or of reason. With this additional claim, it then follows that order implies reason. This supporting argument is shown in Figure 3.2.

---

[29] Hoyle 1983, 19.

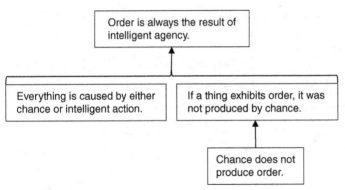

Figure 3.2   A supporting sub-argument in the argument from order

Just prior to the shepherd parable, Balbus gives an example than could be interpreted as a reiteration of the argument from order. He describes an orrery recently constructed by Posidonius of the same sort that Archimedes produced. An orrery – like the Antikythera Mechanism described in Chapter 1 – is a mechanical device that shows the relative positions of the Earth, Moon, and planets. Essentially, it is a model solar system. He claims that, were such a device shown to the 'barbarians' of Scythia or Britain, they would surely infer without hesitation that it was the product of deliberate, intelligent manufacture. It is absurd, says Balbus, to refuse the same conclusion when it comes to the actual solar system, since this obviously has all the features of the orrery, only on a grander scale.

As I said, we could read this argument about the orrery as just another instance of the argument from order: both the orrery and the solar system have particular properties – a sort of order – that cannot result from chance and therefore must derive from intelligent agency. But we could also read this as a distinct sort of argument – the *argument by analogy*. While the argument from order is a deductive inference, at least in its final stage (the top of Figure 3.1), an argument from analogy is a different sort of inference altogether. We'll examine the nature of analogies in more detail in the next chapter. Briefly, this sort of inference works as follows. One notes that a particular item of interest – called the *model* – is similar in many respects to another item called the *target*. One then infers by analogy that some additional property possessed by the model is also possessed by the target. So in this case, we might say that the orrery (the model) is like the solar system (the target) in the relative motion of its parts,

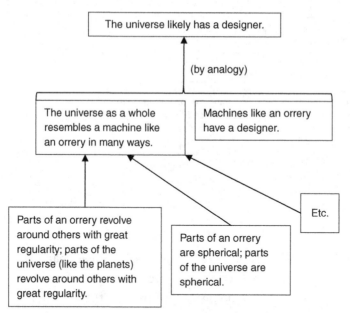

Figure 3.3   The argument by analogy

in the symmetry of its structure, etc. Since the orrery has a designer, we can then infer by analogy that the solar system probably has one too. This is the argument by analogy, represented as a diagram in Figure 3.3. Note that, like inductive arguments in general, any argument by analogy is at best a probable inference; the conclusion does not follow with certainty. The more the model is like the target, the more confident we are of the conclusion. But we can never be sure. We'll see a lot more of the argument by analogy in Chapter 7.

A third argument that one can plausibly extract from the Stoic exposition of Balbus is essentially the argument of Socrates mentioned above. We'll call it the *argument from purpose*. After detailing the intricate order to be found in the heavens (and espousing the argument from order treated above), Balbus turns to consider living things in II.120–32. To give you an idea how this argument runs, here are the first few sentences:[30]

> Now let us turn from the things of heaven to those of earth: is there a single one among them in which the design of an intelligent nature is not obvious? Take for a start the stems of plants which spring from the earth.

[30] II.120, see Cicero 1998, 90.

They both lend stability to the produce which they sustain, and they extract from the earth the sap for the plants which are held in position by their roots. Moreover, they are shielded by bark or cork to give them better protection against cold and heat. Then again, vines cling to their supports with tendrils which serve as hands, and in this way they raise themselves upwards as though they are living creatures.

Each of the biological features described is asserted to serve an obvious purpose, all contributing to the perpetuation of the individual or kind. The upshot is that "[w]e are enabled to understand that none of this happens by chance, and that all is achieved by the provision and genius of nature."[31] All of these examples are understood by Balbus to be obvious instances of contrivance.

Now, I should point out that Balbus does not explicitly argue from the purposeful adaptation of organisms to the existence of god. He thinks that the question of god's existence was dispatched quite early in his discourse with a cursory version of the argument from order and a handful of other arguments such as an appeal to the general consensus of people. At this stage in the dialogue, he is attempting to demonstrate that not only was the universe created by an intelligent god, but that god cares for the well-being of humans. Nonetheless, the intellectual heirs of Cicero will use exactly the premises I've given here in an argument for the existence of god. To do so, one need only complete the argument as Socrates did much earlier by noting that purpose entails design – nothing works toward an end unless it is pushed in that direction by an intelligent being. Thus, the clear purpose for which living things were endowed with adaptations (namely the preservation of myriad forms of life) implies the existence of a designer who had in mind just such a goal. Since the Stoics were consciously borrowing from Socrates, we can turn the argument for god's concern with human welfare into an argument for god's existence without doing violence to either the stated arguments or presumed intentions of the Stoics. The resulting argument from purpose is shown in Figure 3.4.

We can similarly distinguish one final argument in the exposition of Balbus – *the argument from providence* – that appeals to the providential arrangement of the universe for the perpetuation of life and for the

---

[31] II.128, see Cicero 1998, 94.

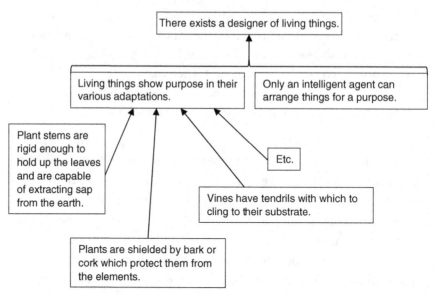

Figure 3.4   The argument from purpose

benefit of mankind. He goes on at great length to provide many examples of the ways in which organisms are adapted to perpetuate the web of life. He also expounds on the ways in which humans are perfectly equipped to make use of their environment and of ways the world is arranged for our benefit and enjoyment. Again, Balbus does not present evidence of divine providence explicitly as an argument for the existence of the gods; he takes the fact of their existence to have been previously established, and claims to be arguing that the gods play an active role in governing the universe. But this also amounts to an additional argument in favor of the existence of the gods – evidence of providential governance allows us to infer design, which then entails a designer. This is how the Academic character, Cotta, seems to take the argument in his reply to Balbus (see Book III of *De natura deorum*). So let's see how Balbus' presentation might be read as an argument for the existence of a god or gods.

The case for providence is made in roughly three parts. The first actually involves more discussion of the arrangement of heavenly bodies, and repeats a great deal of the premises of the argument from order. We'll skip that part. In the second part, Balbus points to the long list of biological adaptations that I have already suggested serve as the basis for an argument from purpose. In this case, these examples are supposed to lend

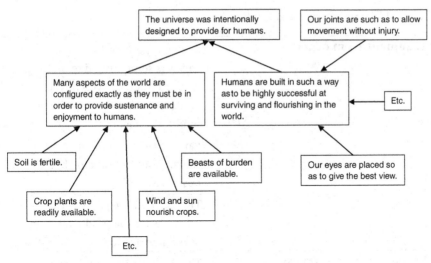

Figure 3.5   The argument from providence

weight to the conclusion that "Divine providence has taken great pains to ensure that the universe is enduringly embellished,"[32] that is, that the universe is perpetually full of living things.

The final part of the argument from providence rests on many examples of the ways in which the universe is apparently tuned or configured for the express enjoyment and maintenance of humans in particular. He cites the abundance and variety of food, the fertility of the ground and usefulness of the trade winds that help us grow the food, and the complex structure of the organs that allow us to digest it. Again, I won't reproduce the entire list, but rather note the conclusion Balbus draws: "the universe and all within it have been created for the benefit of the gods and the human race."[33] The argument from providence is diagramed in Figure 3.5.

These, then, are the principal Stoic design arguments. A short summary is provided in Table 3.1 for future reference. I have left out of the discussion many smaller supporting or tangential arguments. Some of these will make an appearance when we discuss Hume's criticism of design arguments in Chapter 5. For now, let me just point out that the Stoics did not conclude their design argument in the way you might expect.

[32] II.127, see Cicero 1998, 93.   [33] II.133, see Cicero 1998, 96.

Table 3.1 The Stoic design arguments

Argument from order	Some property X (for the Stoics, this was something like mathematical regularity) is always or almost always the product of intelligent agency.
	Parts of the universe such as the motions of heavenly bodies possess X.
	Therefore, those parts of the universe are the products of intelligent agency.
Argument from purpose	The forms of living things (or other aspects of the natural world) show evident purpose (e.g. the purpose of the eye is obviously to see).
	Purpose implies a designer.
	Therefore, there exists a designer of living things.
Argument from providence	Many aspects of the world are configured precisely as they must be for the continued existence, flourishing, and enjoyment of human beings.
	Therefore, the universe was designed for the benefit of humans.
	Thus, there exists a designer of the universe.
Argument by analogy	The universe as a whole in many ways resembles a machine.
	Machines have designers.
	Therefore, the universe likely has a designer.

3.5 The Stoic cosmology

Unlike later Christian writers like Thomas Aquinas, the Stoics do not conclude that an Abrahamic God[34] must be the intelligence behind the world's apparent design. Rather, they conclude that the universe *is* God. Stranger still, they posit that various parts of the universe, such as the heavenly bodies (e.g. stars and planets) are themselves gods – powerful beings possessed of life and reason. I suggest reading an account such as that to be found in Copleston 1993, Vol. I, if you want to make sense of the whole picture, but the idea is roughly this. The universe is all one material thing with two different aspects. On the one hand, there is matter, the passive stuff of which everything is made. On the other hand there is Fire or *pneuma*, the active principle that both creates the lesser elements from matter and shapes the world. This primal Fire is to be identified with God's consciousness. Since the universe is all one thing, God is material. Since God is conscious, the universe is conscious. The point is that, for the Stoics, evidence of intelligence in the universe is not like the evidence of intelligence in an artifact. In the latter case, the creator is separate from the thing created. Instead, the Stoics take such things as order in the universe to indicate that the universe itself is intelligent!

3.6 What's missing?

Many aspects of the Stoic arguments have survived for an enormous amount of time beyond the Stoic school itself. As I suggested in Chapter 1, the motivating intuitions are quite durable. What I called intuitions of form motivate the argument from order and the argument by analogy. Obviously, a perception of purpose plays a role in establishing the premises of the argument from purpose, and a sense of cosmic conspiracy no doubt prompted the argument from providence. All of these intuitions will continue to underlie design arguments for the foreseeable future. But the Stoics gave us more than vague intuitions; they gave us arguments. Unlike intuitions, arguments can secure a contentious conclusion on objective grounds. Each of the four formal arguments presented in this chapter persisted for many centuries beyond the fall of the Roman empire. During that time, each was elaborated, criticized, and adjusted. If we want

---

[34] I mean the God of the Old Testament, who exists independently of the universe, his creation.

to know whether the observable world offers evidence for the existence of God, we will have to grapple with these four arguments.

This is not to say that the Stoics gave us the strongest possible form of each argument. There are, as we'll see, many features of the Stoic arguments that are left wanting. There are many questions that need to be addressed. For instance, how plausible is the dichotomous view of causes that sees all events as the outcome of either chance or reason? At the scale of human experience (that which was accessible to the Stoics of the first century BCE) it is perhaps not so implausible. We do not see rocks align themselves in grids or rivers run in straight courses like canals. Only the products of human construction bear such simple geometries when viewed on the scale of meters. But what about now? What does the modern scientific perspective have to say about the occurrence of the order or regularity that so impressed the ancient astronomers? For one, the orbits of planets can be explained by the action of mindless gravity just as the motion of an orrery can be accounted for by the action of its internal clockwork. The atoms in rock crystals and the arms of snowflakes are highly ordered without any obvious intervention by an intelligent agent. These are things to think about as we look at more sophisticated presentations of the argument. As we move on throughout our survey, consider as well the following three questions: (i) why should we believe that order derives only from intelligence? (ii) can one really assert that the world was made for humankind or was humankind in some sense fashioned for the world, perhaps by unintelligent causes? and (iii) how good is the analogy between the universe and such things as clocks or orreries?

4   Medieval arguments

4.1 Introduction

Our aim in scrutinizing the historical literature is to identify and assess the major forms of design argument. In the last chapter, we considered the design arguments available as of about 100 BCE, all of which were defended by the Stoics at one time or another. Now we're going to glide over the next fifteen centuries or so, stopping only to scrutinize a single argument from the High Middle Ages. The reason for our haste is that the development of design arguments largely stagnated from the time of Cicero until the scientific revolution. This wasn't for lack of interest in things divine. Rather, it is a consequence of historical forces that temporarily suspended the broader project of *natural theology* – the attempt to address questions concerning the existence and character of God on the basis of reason and experience.[1] These historical forces center on Rome. In the fourth century CE, the Roman empire in effect converted to Christianity following the Emperor Constantine. As the last Roman emperor to rule over a united empire, Constantine moved the official capital to Byzantium, well east of the empire's traditional seat of power in Italy. Not long after his death, the empire permanently split in two, and the two fragments followed very different intellectual trajectories with rather different perspectives on design arguments.[2] For this reason, the intellectual history of design arguments through the following centuries is best told in two parallel stories, one for the Christian West and one for what would ultimately become the Islamic East.

---

[1] For a modern treatment of the distinction between 'natural' and 'revealed' theology, see, e.g., Hick 1963. For a more thorough overview of the project and aims of natural theology, see, for instance, Valentine 1885.

[2] Sinnigen and Boak 1977.

4.2 Medieval arguments from the Islamic East

The eastern fragment of the Roman empire kept Greek philosophy and the Stoic interest in design arguments alive. Centered in Constantinople, it evolved into what we moderns call the Byzantine empire, which lasted until the fifteenth century CE. Though romanized, spoken Greek persisted there, and this doubtless helped to preserve Hellenic philosophy. It was from the Byzantine world that Greek philosophy made its way into Islamic culture. In the mid seventh century CE Arab armies began spreading Islam by force. In 641 CE they captured the city of Alexandria, whose vibrant intellectual life had survived the collapse of the Roman empire. This city's rich collection of Greek and Roman philosophy – most prominently the works of Plato and Aristotle – thus found their way into Arabic translation, and provided a foundation for the development of Islamic philosophy.[3]

Unlike their Christian counterparts of the time, early Islamic philosophers retained an interest in natural theology. However, their emphasis was on deductive argumentation and in particular on the so-called 'cosmological argument' derived from Aristotle's notion of an unchanging source of change (the so-called 'first-mover'). Design arguments were kept alive but remained largely in their Stoic form. Deemed less rigorous than the cosmological argument, design arguments were seldom closely reasoned and were mostly used for purposes of popularization.[4] For instance, Averroes (1126–98 CE) – a giant of Islamic philosophy and an important commentator on Aristotle[5] – provided only a loose rendition of what in Chapter 3 we called the argument from providence:[6]

> "Everything in the world is adapted" to the needs of the human species and reveals "providence." "Day and night, sun and moon," the earth and everything therein, the organs of the human body – all serve the needs of man. The functionality exhibited throughout the world cannot conceivably be due to "chance." It must "perforce" be the doing of "an agent … who intends … and wills it"; and the "existence of a creator" is thereby established.

[3] Esposito 1999, 271.   [4] See Davidson 1987, ch. 7.
[5] Hyman and Walsh 1973.
[6] The quoted text is a reconstruction of Averroes' argument given in Davidson 1987, 229.

The story of design arguments in the East reunites with that of the West around the twelfth century when the Arabic translations of ancient philosophical works long lost to Christian scholars were reintroduced to Western philosophers. Though Islamic philosophers offered little innovation in design arguments beyond the Stoic arguments, they did develop the project of natural theology. Their approach – anchored in the work of Aristotle – had an enormous influence on Western philosophers like Thomas Aquinas.

4.3 Medieval arguments from the Christian West

In the West, our story begins with an ending. It took only a century and a half after the passing of Constantine for the Western fragment of the old Roman empire to dissolve into a bewildering patchwork of Germanic kingdoms, with an Ostrogoth named Theodoric ruling Rome as king.[7] As Roman political control waned, the influence of the Church that Constantine had embraced waxed. In part, this was because the Germanic tribes picking over the bones of the Roman empire were largely Christian. Thus, the fall of the Roman empire in the West ushered in the rise of the Roman Catholic Church as both a religious and secular authority.

For a little while, philosophy survived this change in masters. However, under the influence of the Church it was yoked to the project of clarifying, defending, and justifying Christian doctrine. Two of the most important figures in this tradition – neither of whom bequeathed anything like a design argument – were Augustine (354–430 CE) and Boethius (c. 476–c. 526). Augustine, like most of the early Church Fathers, was heavily influenced by Plato, particularly in his penchant for elaborating visions of the world rather than arguing for principles and in his general view that the material world is little more than a corrupting morass in which the soul is temporarily stuck. Augustine's Platonic approach to philosophy characterized Christian thought for centuries. Boethius was a tragic figure who wrote one of philosophy's most poignant works, *The Consolation of Philosophy*, as he awaited his execution in prison. Perhaps more important for the history of philosophy, Boethius translated the *Organon*, Aristotle's collected works on logic, from Greek – which was a dying language in the

---

[7] Sinnigen and Boak 1977.

West – into Latin. Though he never got around to the rest of Aristotle's works as he had intended, his translation of the *Organon* made almost the entirety of Aristotle's work available to the West for centuries – a dim light during the Dark Ages.[8]

Not long after the death of Boethius, the project of the Church Fathers ground to a halt and philosophy in the West was all but extinguished for the next few centuries. In the twelfth century, however, Aristotle made a dramatic return as Christian scholars translated works that had been kept in circulation by Islamic philosophers. This rediscovery of classical learning reignited philosophy in the West. At the forefront of the philosophical revival was Thomas Aquinas, the author of the only significant medieval design argument. Aquinas managed the Herculean task of producing a synthesis of Aristotelian philosophy and Christian theology. This synthesis, however, began to unravel under the scrutiny of William of Occam in the early fourteenth century, and hope faded once again for the project of natural theology. That is, until the scientific revolution. As natural science began to emerge in the sixteenth century, the old arguments of the Stoics gained new life. This is where our survey will resume in the next chapter.

4.4 Thomas Aquinas

As I mentioned above, Thomas Aquinas produced the only notable design argument of the Middle Ages. In order to give this argument a charitable hearing, it will help to situate him historically. Aquinas was born to a noble family[9] at the end of 1224 or beginning of 1225.[10] He began his education at the Benedictine abbey of Monte Cassino, and in his early teens was sent to the University of Naples to study the liberal arts. To get a sense of the changing intellectual climate in which he lived, it helps to note that the University of Naples – one of the first universities ever – was founded

---

[8] It's a curious fact that Cicero's *De natura deorum* remained in uninterrupted circulation in the West (Cicero 1998, xlviiii).

[9] Thomas' forebears were the counts of Aquino, hence the name Thomas d'Aquino, or Thomas Aquinas.

[10] For a brief sketch of Aquinas' historical context, including the biographical details offered in this paragraph, see Aertsen 1993. For a more thorough treatment of these topics, see Weisheipl 1983.

in the year of Aquinas' birth. After Aquinas joined the Dominicans – one of the newly sanctioned mendicant orders of friars – he was dispatched to Paris to study with Albertus Magnus (Albert the Great). Aquinas stayed on as master of theology in Paris for a few years, then taught in various cities. All the while, he was deeply involved in the disputes between the Faculty of Arts, which studied secular philosophy and science, and the Faculty of Theology, which largely occupied itself with training clergy. Thomas Aquinas died in 1274 at the relatively early age of 49, and was ultimately pronounced a saint by the Catholic Church.

Most important for understanding his philosophical work is the fact that Aquinas wrote – and arguably thought – in the 'scholastic method'. This was the dominant literary format of medieval philosophy and imitated the highly structured classroom disputations that comprised much of medieval university instruction. In the scholastic method, treatises on various subjects are rigidly divided into 'questions' corresponding to major topics. These are further divided into 'articles' which provide detailed examinations of different positions with respect to the question being considered. The examination of individual theses within an article proceeds according to a prescribed structure as well.[11] To the modern reader, the rigid structure of scholastic writing tends to seem overly pedantic and inaccessible. While there is some truth to this impression – one needn't spend much time reading Aquinas' examination of the substance of angels to despair of medieval philosophy – it is also largely unfair. There is a utility in the form of scholastic writing. According to my favorite characterization of scholastic philosophy:[12]

> [S]cholastic philosophy is hard and dry for much the same reason as
> a beetle is hard and dry: its skeleton is on the outside. Argument,
> the skeleton of all philosophy, has been on the inside during most of
> philosophy's history: covered by artful conversation in Plato, by masterful
> rhetoric in Augustine, by deceptively plain speaking in the British
> empiricists ... And the scholastic method – laying out the arguments
> plainly and developing the issues in such a way that both sides are
> attacked and defended – provides an opportunity, unique among the types
> of philosophical literature, for understanding the nature of philosophical
> reasoning and assessing its success or failure.

[11] See Aertsen 1993.   [12] Kretzmann and Stump 1993, 6.

As described here, the scholastic method, irrespective of the content of medieval arguments, provides us with a model of how an inquiry ought to be conducted. Our goal in this text is to strip away the rhetorical meat of any design argument and lay bare its skeleton – to turn it inside out and put the skeleton on display. In the scholastic spirit, every argument we consider is vetted by attacking and defending it as best we can before taking the balance.

4.5 Aquinas' Fifth Way

The only medieval design argument we'll scrutinize in this way appears early in Aquinas' *Summa theologica*, a hefty three-part tome dealing with theology along with a host of other topics including metaphysics, ethics, and law. Being a masterpiece of scholastic writing, the work is organized around questions, the second of which concerns the existence of God. In the third article under this question, Aquinas offers five arguments for the existence of God, known to us moderns as the 'Five Ways'. The famous Fifth Way is our first example of a design argument. Here is how Aquinas puts it:[13]

> The fifth way is taken from the governance of the world. We see that things which lack intelligence, such as natural bodies, act for an end, and this is evident from their acting always, or nearly always, in the same way, so as to obtain the best result. Hence it is plain that not fortuitously, but designedly, do they achieve their end. Now whatever lacks intelligence cannot move towards an end, unless it be directed by some being endowed with knowledge and intelligence; as the arrow is shot to its mark by the archer. Therefore some intelligent being exists by whom all natural things are directed to their end; and this being we call God.

To make analysis of the argument a little easier, we'll consider instead a modern reconstruction of what Aquinas is saying:[14]

> We observe that some things which lack awareness, namely natural bodies, act for the sake of an end. This is clear because they always or commonly act in the same manner to achieve what is best, which shows that they reach their goal not by chance but because they tend towards it. Now things which lack awareness do not tend towards a goal unless

---

[13] Aquinas 1920, 14.   [14] Kenny 1969, 96.

directed by something with awareness and intelligence, like an arrow by an archer. Therefore there is some intelligent being by whom everything in nature is directed to a goal, and this we call 'God'.

This latter version of the argument can be diagramed as in Figure 4.1. In diagramming the argument, I have labeled the statements for easy reference. Those marked with a 'P' such as (P1) are premises, the conclusion is indicated by a 'C', and sub-conclusions are number with an 'SC' prefix. Note that I have also included an *implicit premise* – a premise that the author does not explicitly state but which we can assume he intended since it is needed for the inference to work. It is labeled (IP1) in the diagram. With the argument laid out in this way, there are two features to note at the outset. First, it is plainly a posteriori – the argument appeals to experience in a big way when it says "we observe that some things ..." Second, the Fifth Way is a deductive argument – if the premises are true, the conclusion is supposed to follow with certainty. This is true for each of the individual inferences (represented by arrows) that make up the argument, and thus for the argument overall. So in this case, in order to decide whether the argument establishes the conclusion, we first need to ascertain whether each inference – and thus the argument as a whole – is valid.

We are immediately faced with a problem. Both Aquinas' original argument and the modern reconstruction are ambiguous with respect to (P1) and thus (SC1). One obvious approach is to understand (P1) as the claim that inanimate objects *individually* act in the same manner – different for each object – to achieve what is 'best' relative to their respective situations. So for instance, a dropped tennis ball tends toward the ground while the sun tends toward motion in a circle about the Earth (at least from Aquinas' perspective). Furthermore, experience tells us that each object acts consistently in a given context, but each object acts differently in different contexts. For instance, an arrow fired into the air always follows a parabolic path, but an arrow launched underwater follows an Aristotelian trajectory, quickly coming to a stop before drifting straight downward. In the most natural reading, then, we should understand (P1) as the claim that natural objects consistently tend toward the same outcome or state of affairs in any given context (we'll grant for the sake of argument that this outcome is 'best'). If we read (P1) in this way, the sub-conclusion (SC1) must be that natural bodies each act toward some *individual* goal. But now there is a problem. The inference from (SC1) and (P2) to the conclusion that there

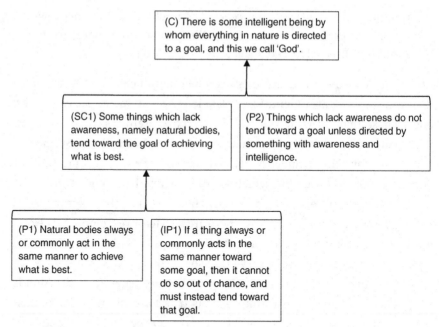

Figure 4.1   Aquinas' Fifth Way

exists a single intelligent being directing all natural bodies is invalid. The inference has the following form:

> Premise: For all *x* there exists a *y* such that *x*R*y*.
>
> Conclusion: There exists a *y* such that for all *x*, it is the case that *x*R*y*.

The notation *x*R*y* is just shorthand indicating that some relation R holds between *x* and *y*. In this case, you should read *x*R*y* as '*x* is directed by *y*'. This inference form is clearly invalid. In fact, this mistake has a name. It is known as the 'quantifier shift fallacy'. It is easy to see what has gone wrong if we consider an example Elliott Sober calls the Birthday Fallacy.[15] It goes like this:

> Premise: For every person *x* in this room, there exists a day *y* such that *x* was born on *y*. (For all people in the room, there is a day on which that person was born.)
>
> Conclusion: There exists a day *y* such that for every person *x* in this room, *x* was born on *y*. (There is a day on which every person in this room was born.)

---

[15] Sober 2005a.

Obviously this is absurd. Just because we each have a birthday does not
mean that we all share the *same* birthday. Likewise, just because every
natural body is guided by an intelligence it does not follow that every natural body is guided by one and the same intelligence. Aquinas' argument
in this reading is invalid insofar as it purports to establish the existence
of a unique God.

We can get rid of the quantifier shift fallacy if we read premise (P1)
differently and simultaneously strengthen premise (P2). Suppose (P1) is
instead the claim that all natural bodies tend towards the *same* end. In
other words, suppose that we take it for granted that even though they
have distinct tendencies, every natural body actually contributes to the
attainment of the same state of affairs for the world – though each object
does its own thing, in doing so the whole observable world tends toward
a predetermined outcome. There is precedent for this idea. Aristotle, for
example, held that all earthy objects tend toward the center of the universe. Even though each object falls or rolls in a different direction locally,
all objects jointly tend toward a state of affairs in which the earthy objects
of the world are concentrated at the center. Other precedents can be found
in earlier Jewish philosophy, where for instance Bahya ibn Paquda noted
a "single order and a single motion, which comprehends each and every
part" of creation.[16] In modern physics, we talk of systems of objects tend-
ing toward the state of lowest energy or maximum entropy. So it is at
least plausible to reinterpret (P1) in such a way. We can also make (P2) the
claim that inanimate things tend toward a common goal only if they are
directed by a single intelligence. If we make both these changes, then the
argument is valid. The repaired argument is shown in Figure 4.2. I have
added another implicit premise (IP1) to help make the inference clear.

Of course, validity is only half the story. We want to know whether
the conclusion of the argument is true. Recall that, for a deductive argument to support the truth of a claim, the argument must be sound. To
be sound, the argument must be valid and all premises must be true.
Here our strengthened premise (P2) does not fare so well. It is simply false
that the coordinated motion of inanimate objects can only be produced
by single intelligence directing the whole show. We know, for instance,
that the components of a house – each of which lacks intelligence – can

---

[16] Davidson 1987, 228.

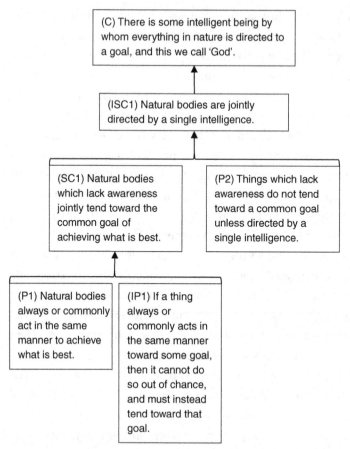

Figure 4.2 A modified version of Aquinas' Fifth Way contrived to avoid the quantifier shift fallacy

tend toward the common end of assembling a home while being guided by *many* human intelligences, not just one. The strengthened version of (P2) is plainly false.

I won't attempt a second round of revision, but our first attempt illustrates an important feature of arguments: they are integrated wholes. A change in one inference or one premise ripples throughout an argument, affecting the validity or strength of other connected inferences. In this case, if we fix the false premise it is not clear that we can maintain a valid argument in favor of a unique God. Of course, even if we do there are other questionable premises. How do we know what is 'best'? How do we know that coordinated tendencies suggest intelligence unless we can identify a goal of the particular intelligence in question? These concerns will arise again when we consider later design arguments that appeal to the purposeful operation of the inanimate world, and so I'll put off further discussion until Chapter 10.

Aquinas produced the most closely argued if not the most original medieval design argument. It is a posteriori in that it appeals to experience in order to establish that inanimate things act towards a goal. It is deductive in that its conclusion is supposed to follow with certainty. By casting the argument in the form of a diagram, we were able to quickly isolate ambiguities and to repair a logical fallacy, though even the repaired argument is of dubious soundness. In the next chapter, we will apply these same analytic tools to the new breed of design argument ushered in by the scientific revolution.

5   The golden age of natural theology

5.1 The rise of natural theology

From among the sparse offerings of the Middle Ages, I presented only the design argument of Thomas Aquinas as worth a close examination. However, there is one more medieval work we ought to consider, if only for its significance as a transitional text. I am referring to a book written in the 1430s by a Spanish professor of medicine, philosophy, and theology named Raymond Sebonde. The importance of this book is not to be found in the specific design argument it contains. In fact, unlike what we saw in the work of Aquinas, it's no straightforward affair to extract a clear argument from Sebonde. His book was never translated from its original Latin into English (though Montaigne did translate it into French), and those few who have undertaken to study the book in its original form suggest it is almost unreadable. It contains some 330 chapters of poorly rendered and heavily abbreviated Latin. Nonetheless, we can charitably reconstruct the following argument from the third chapter of the text.[1] The "book of nature" is composed of creatures that can be sorted into four "degrees" or "grades" based on their possession or lack of "existence, life, sensation, intelligence, and free will." In the first grade are things like rocks that have only existence. In the second grade are things like plants that have existence and life, but nothing else. The animals, the third grade of creature, have everything excepting intelligence and free will. Mankind is in a grade of its own, possessing the whole package of characteristics. Now, says Sebonde, the presence of these characteristics must have a cause. The cause cannot be humanity itself – we certainly didn't give ourselves

---

[1] Both for what details are known of Sebonde's life and for an English translation of the third chapter, I have relied on Hicks 1883. Hicks also presents a slightly different but entirely plausible reading of the argument.

intelligence or free will – or any of the lower grades of creatures. Thus, the cause must be something with more numerous and more powerful characteristics than humans. Furthermore, the hierarchical order in the grades of creatures – man above animals above plants above minerals – could only result from the intention of an intelligent creator. Lastly, we know there's only one creator since the characteristics are the same for each grade in which they appear and therefore must have the same cause. That intelligent, powerful cause is God.

It may be obvious by this point why I skipped this argument in the last chapter. The part that concerns the existence of God amounts to an argument from order that rests on a division of things that is arguably arbitrary. That is, the order being singled out is plausibly an order imposed by Sebonde, not a feature of the world. We might just as well have pointed out that some things are only blue while others are blue and round, and use this hierarchical structure to argue for a creator. In fact, any random collection of particulars can always be sorted in such a way. Thus, much more would have to be said about the hierarchy of existence before we recognize it as possessing an order that could only derive from intelligence. I won't pursue this line of criticism any further. Whatever Sebonde intended and whatever the merits of his idiosyncratic argument, Sebonde's book is important for what it presaged. The title of the work was *Natural Theology*, a phrase invented by Sebonde, and it represented a radical departure in attitude from the medieval schoolmen like Aquinas. Sebonde gave a posteriori arguments a central role. However muddled his arguments, he set out to demonstrate the existence of God solely on the basis of the form and contents of the natural world. Both this emphasis on a posteriori argument and the name of 'natural theology' which Sebonde gave it stuck. The explosion of arguments for the existence and attributes of God that began in the mid seventeenth century were predominantly of this type. So with a nod to Sebonde, we can accurately characterize the period from about 1650 to 1850 as the golden age of natural theology.

It's an interesting question why there should be a golden age at all. What could have precipitated such an interest in design arguments? As I mentioned in the last chapter, there weren't very many explicit arguments for the existence of God offered during the medieval period, and even fewer that could be construed as design arguments. I suggested that this was because there was no one to defend (publicly at least) the contrary position, i.e. that God does not exist. In this respect, not much had changed by the dawn of the seventeenth century when the flood Sebonde started began to crest. With the possible exception of Thomas Hobbes, atheists were hard to find in print. In fact, it wasn't until 1782 that the first explicit argument against the existence of God appeared in print.[2] So who were all these natural theological arguments directed at? Some historians argue that the sheer volume of polemic against the atheists is evidence enough that even though they didn't publish, there must have been plenty of atheists lurking about in salons (of the French intellectual variety).[3] Others have argued that the intellectual, literary, and judicial assault on atheism was more repressive than oppressive – intended to discourage anyone from developing an atheist position rather than an attempt to quash an existing philosophy.

Of course, I've been using 'atheist' in the modern sense of one who denies the existence of God. The epithet 'atheist' was applied rather more liberally by seventeenth- and eighteenth-century authors so as to include more or less anyone who expressed disbelief or doubt in the nature and character of God as set forth in Protestant (or sometimes Catholic) doctrine. In this sense, there were a number of prominent public 'atheists', many of whom were proponents of something called 'deism'. Roughly, deism is a family of views that rejects the possibility of revelation and the Christian view of an immanent God who cares about and interferes in the goings-on of the universe. The deists believed in an impersonal God, often cast as a clockmaker who set the mechanism of the universe going and stepped away. But the deists did accept the existence of a God, albeit a rather impersonal one. So it seems that design arguments would gain little purchase on their position. We are left, then, with a bit of a puzzle.

Whatever the cause, there is an undeniable pattern. Wherever the new and growing natural sciences mixed with Protestantism, the result was a profusion of natural theological arguments, particularly design arguments. In fact, much of the early scientific literature in biology and geology was difficult to separate from the literature on natural theology. A great many scientists, some of whom we'll consider below, saw themselves serving both the purposes of science and theology in their work. We haven't space to consider all that was said on the subject (that would take an encyclopedia), but we can identify the major distinct species of

[2] See Berman 1988, ch. 5.   [3] Popkin 1992.

design argument advanced in this period. This chapter and the next look at the types of design argument in play by the early eighteenth century. Chapters 6–9 will cover the rest of the golden age, ending with the publication of Darwin's *Origin of Species* in the 1860s.

5.2 Boyle

Any discussion of natural theology in the seventeenth century or beyond would be incomplete without mentioning Robert Boyle. He was, of course, a prominent figure in the emergence of modern science – remembered as much for founding the Royal Society as for the law concerning gas pressure that bears his name. More to the point, he was a conspicuous and ardent supporter of natural theology and a proponent of at least one design argument. So great a supporter was Boyle that in his will he bequeathed funds for a memorial lecture series aimed at "proving the Christian religion against notorious infidels, viz. Atheists, Deists, Pagans, Jews, and Mahometans."[4] His wishes were carried out, and a parade of notable thinkers, many deeply engaged in the project of science, successively spoke on the inference from nature to God. Whether the lectures succeeded in Boyle's stated aim is subject to debate – an oft-repeated adage says that no one doubted God's existence until Boyle set out to prove it. But it is less controversial that Boyle's lectures reinvigorated the ancient debate between proponents and critics of design.

A deeply pious man, Boyle's scientific and naturalistic papers are saturated with declarations about the glory of God revealed by the study of nature. For him, there was no question of God's role in the design and creation of the cosmos. While he often gestured at it, he seldom if ever explicitly articulated a design argument. Nonetheless, such an argument is implicit in a number of his works. A clear example can be found in *A Disquisition about the Final Causes of Natural Things*. In this text, Boyle is concerned with the problem of identifying ends or goals in the works of nature. He is not concerned with proving that there exists a creator God whose intentions determine the ends of natural phenomena – he takes this for granted – but rather whether we can ever identify those ends in particular cases. Contrary to Descartes and his followers, who think it impossible for

[4] Bentley 1966, xv.

we limited human beings to know any of the goals of an infinitely powerful and intelligent God, Boyle thinks we can identify some of these goals in the operation of nature. Importantly, he seems to think that we can identify some goals or 'final ends' without knowing much of anything about God. Of course, if we have a means of identifying the presence of an intentional end – of a purpose for which an intelligent agent produced a given object or state of affairs – without knowing in advance that such an agent even exists, then we would have the essential components of an effective design argument. When considering living things, Boyle thinks we have just such a case. While he concedes that random processes can produce impressive outcomes such as mineral deposits shaped like fern leaves or patterns in marble that resemble pictures of towns or men, he asserts that some functions performed by parts of living things involve[5]

> such a number and concourse of conspiring causes, and such a continued series of motions and operations, that 'tis utterly improbable, they should be produced without the superintendency of a Rational Agent, Wise and Powerfull enough to range and dispose the several intervening Agent's [sic] and Instruments, after the manner requisite to the production of such a remote effect. And therefore it will not follow, that if Chance could produce a slight contexture in a few parts of matter; we may safely conclude it able to produce so exquisite and admirable a Contrivance, as that of the Body of an Animal.

Stated with a little more care, here is the argument that emerges from Boyle's discussion:

(1) Some natural phenomena, particularly the structure and physiology of living things, involve a complex chain of interacting parts that results in the performance of a function (such as vision) which in turn contributes to the persistence and reproduction of the organism.
(2) Such a complex, finely configured causal chain in which each part has precisely the properties it needs to contribute to the function of the chain is overwhelmingly unlikely to have been produced by chance.
(3) There is no reason to reject the persistence of living things as an end of an intelligent creator.
(4) Therefore, there is a God who created living things.

---

[5] Boyle 1688, 45–46.

This argument is a hybrid of the Stoic arguments from order and purpose. The basic strategy is to identify instances of purpose. To do so, we appeal to the inability of chance to produce a complex outcome, just as in the argument from order. The question, then, is whether Boyle's argument is an improvement over the Stoics'. The answer depends in large measure on the truth of an implicit premise needed to make the argument valid: the agency of an intelligent being is the default explanation when chance is ruled out. This isn't quite the same as the Stoic claim that the only two possibilities are chance and design – Boyle's argument leaves open the possibility of other causes. Of course, Boyle the man is already convinced that God is responsible for everything. But his argument does allow for the possibility of rejecting intelligent (divine) agency as an explanation if we think that the function of the complex causal chain in question cannot plausibly be a goal of an intelligent creator. How do we know what a plausible goal is? Boyle seems to take it as obvious that a plausible goal is anything a loving God like that of the Christian (Protestant) faith would be interested in producing.

The upshot is that Boyle admits at least the possibility of other causes besides chance and intelligent agency. The problem, however, is that Boyle's method is heavily weighted in favor of intelligence irrespective of the evidence. This is a problem faced by any default-favoring inductive principle – a rule for inductive inference by which a default explanation is accepted unless eliminated by the evidence. Since very strong evidence is required to rule out the default explanation – it is remarkably hard to determine that a particular outcome could not be the aim of an infinitely wise creator – then the method implicitly puts great confidence in the default explanation a priori. In this case, that means that the hypothesis of an intelligent creator is assumed to be probably the case before we even look at the evidence. For Boyle, there is no conflict here – he already believes that all things living or not were created by God. But if the point is to provide an argument from the evidence of experience to the existence of one or more intelligent creators, then we beg the question by assuming there probably exist one or more intelligent creators.

5.3 Bentley and Newton

As I mentioned above, Boyle bequeathed a sum of money upon his death to endow a series of lectures on natural theology. The very first person to be tasked with defending Christianity from the 'atheists' with the sword of natural science was Richard Bentley. Bentley, born the son of a well-off yeoman in 1662, was just about 30 when the trustees of Boyle's lecture-ship nominated him. Bentley had already made a name for himself as a philologist, though his greatest successes still lay ahead of him. Given that Bentley was not a scientist his nomination was something of a long shot.[6] Yet the trustees could hardly have hoped for a better expositor.

Bentley gave a total of eight lectures (which he calls sermons), spanning an extraordinary range of design arguments. In the first five sermons, Bentley provides arguments that appeal to the faculties of the soul and to the adaptedness of human anatomy. We will focus on the last three of Bentley's lectures, which provide arguments based on the structure of the cosmos and the nature of the laws governing it, in particular the law of universal gravitation. While the overall argument strategy was Bentley's, key components were provided by Sir Isaac Newton. Newton, of course, is what we now call a 'household name', and rightly so – he made foun-dational contributions to mechanics, astronomy, gravitation, optics, fluid dynamics, and mathematics. When Bentley set about composing his Boyle lectures in the early 1690s, Newton was perhaps already the most eminent figure in natural science. It says something about both Bentley's thorough-ness and his ego that he sought Newton's counsel. Judging by the letters preserved by Bentley, Newton was eager to aid the project of natural theol-ogy and played a significant part in shaping this portion of Bentley's argu-ments. Newton himself published a few short design arguments, such as that which appears in the General Scholium of his *Principia*. But Newton's argument is effectively contained within Bentley's extended treatment, so we can focus on the latter.

What is Bentley's argument from "the origin and frame of the world"? It begins with chaos. In previous sermons, Bentley had already argued that the present system of the world could not be eternal, and that in general matter in motion could not have persisted from eternity. But supposing this is not the case, supposing for the sake of argument that matter in motion might have existed from eternity, Bentley aims to demonstrate that this cannot be sufficient to account for the order we currently see in the layout of the cosmos. That is, Bentley will show that a material world

---

[6] See Miller 1978.

beginning without order or structure will never produce it. As with the old Stoic argument from order, Bentley proceeds by providing an exhaustive list of all the processes that could be operative in the world and eliminates all but intelligent design as the source of the world's cosmological order. Like Boyle, he embraces three possibilities: chance, production by some sort of inherent mechanism (mechanical), and intelligent agency. However, Bentley argues that what we really mean when we attribute an outcome to chance is that the outcome was produced in some sort of mechanical process in which intelligence played no role. In essence, saying that something happened by chance is just to say that no one intended it to come out that way. Bentley collapses the notion of chance and production by mechanism into the same thing: "So that this negation of consciousness being all that the notion of chance can add to that of mechanism, we, that do not dispute this matter with the Atheists, nor believe that atoms ever acted by counsel and thought, may have leave to consider the several names of *fortune*, and *chance*, and *nature*, and *mechanism*, as one and the same hypothesis."[7] His task, then, is to show that there is no mechanism, no lawlike process, that could produce our ordered universe from a state of chaos. He proceeds in two steps: (i) characterizing the state of chaos, and (ii) rejecting each possible mechanism in turn.

With respect to the initial state of chaos, Bentley performs an interesting calculation – he computes the volume of empty space that would, on average, surround a particle of matter if the particles currently making up the massive bodies of the world were spread more or less evenly over all of space. So far as I know, he is the first to undertake such a calculation. I won't go into the details of the calculation (rather fascinating in its own right), but the answer is striking: "the empty space of our solar region ... is 8,575 hundred thousand million million times more ample than all the corporeal substance in it."[8] If anything, his estimate is conservative. If we treat the solar system as the sphere enclosing half the distance from Earth to the nearest star, then the ratio of void to the volume of the typical atom (it's pretty much all hydrogen in here) is about a thousand times greater than Bentley's figure.

Whatever the precise value, it's large. And this has important consequences for the ability of mechanical processes to produce order from the

---

[7] Bentley 1966, 148.   [8] Bentley 1966, 153.

void. Bentley says there are really only two candidates. First, we might imagine, as the ancient Greek philosophers Democritus and Epicurus did, that the matter of the world is composed of atoms that move through the void and interact only upon collision. The atoms in this view are not generally spherical, but have all manner of shapes. Some of these are prone to stick together, and so out of the random motion of the atoms, we might slowly accumulate clumps that become stars and planets, etc. Introducing just a bit of contemporary physics, Bentley further supposes that between collisions, atoms move only in straight lines with a speed determined by the "impelling force" they receive from each collision. Now given what was said before about the very low density of material particles, the rate at which they collide in the chaos must be vanishingly small, and so it would at the very least take an enormous amount of time for any structure or order to emerge.[9] More to the point, features of the order we have simply cannot be sustained let alone created in such a process. In particular, Bentley points to the orbits of the planets. Even if something as large as a planet should happen to accrete from the thin chaos of atoms in the void, there would be no way to keep it in a closed orbit. Since the only interaction among particles is mediated by collision, to turn the Earth, say, away from straight line motion and back toward the Sun would require an extraordinary number of particles to collide with it at just the right angle with just enough speed, and to do so continuously from moment to moment. In short, we need a vortex of swirling atoms, such as that proposed by René Descartes in his *Principles of Philosophy*. The only way to sustain such a vortex locally is to fill all of space with moving atoms. But we already know this cannot be the case – the universe is mostly empty space!

Having ruled out the Epicurean mechanism, Bentley has just one left: gravitation. Here he has in mind, of course, the law of universal gravitation as espoused by Newton. His first couple of objections concern the nature of gravity. In particular, Bentley thinks it impossible for gravitational attraction to inhere in objects – gravity cannot be a built-in property of matter. His first argument to this effect begins by noting that, if

---

[9] Bentley seems to offer a fallacious argument as to why the probability of collision must be zero. Read charitably, however, he is merely pointing out that it is very tiny and an enormous amount of time would be required to produce any structure at all.

gravitational attraction is inherent in matter, then there could never have been a state of chaos in the first place. Remember, the material hypothesis for Bentley involves the claim that matter in motion has always existed. If the order we see arose spontaneously at some point, then it must be true that for an infinite duration in the past, matter was scattered about more or less evenly in the universe (this is what Bentley means by chaos). But in that case, matter could not have been gravitating for all that time. Either particles attract gravitationally by their very nature, in which case they would clump together immediately, or they do not attract by nature – there cannot be a time at which it just happens that they start attracting one another.

In his second argument, he suggests that the very phenomenon of gravitational attraction itself amounts to evidence for an intelligent agent behind the order in the universe. To see why, we first have to be clear about what is meant by gravitational attraction. As Bentley puts it, "'tis an operation, or virtue, or influence of distant bodies upon each other through an empty interval, without an *effluvia*, or exhalations, or other corporeal medium to convey and transmit it."[10] In other words, gravity is action at a distance. Since the idea of matter involves only extension, shape, and motion, we can without contradiction imagine matter without gravity. And we cannot explain gravitation by appealing to the intrinsic properties of matter (shape, motion, etc.). Furthermore, we cannot understand gravity as arising from some sort of mechanical interaction between bits of matter that is mediated by other matter (here Bentley differs from Newton). Thus, gravity must be a behavior or property or tendency that is superadded to matter. Such an organizing principle could only come from an intelligent designer of the cosmos.

Perhaps you, as a modern reader, are not convinced by this line of argument. After all, it rests on the assumption that sheer action at a distance through a vacuum is a contradiction in terms. To modern ears, that might sound quaint, though arguably this is merely because we have become accustomed to the idea of action at a distance, not because it makes any more sense to us than it did to Bentley. Fear not, for Bentley has another objection up his sleeve. He'll grant for the sake of argument that a gravitational attraction is an inherent or innate property of matter ('essential' in

[10] Bentley 1966, 163.

his terminology). Then it is still the case that the structure of the world as we see it could not be produced by mechanical means.

The reason has to do with the detailed nature of gravity. As Newton emphasizes in the letters he exchanged with Bentley, there is no way to place the planets in the chaos such that by gravitational attraction alone they achieve fixed orbits. Suppose that, by whatever means, the planets and Sun and other components of the solar system we inhabit formed from the chaos. Is there any way for gravity to bring them together into the system we have? The answer, according to both Bentley and Newton, is negative. To put the argument in somewhat modern terms, the equations governing the motion of massive bodies under the influence of gravity look the same both forward and backward in time. This means that we can determine whether or not a particular process – e.g. the formation of the solar system from randomly spaced planets at rest – is possible by determining whether the reverse process is possible. Since the solar system is stable (more or less), it is not possible under the influence of gravity alone for the planets to suddenly break orbit and fly off to distant locations, at least not without adding energy to the system from without. Thus, we know that there is no particular configuration of planets at rest or in motion that would result in the solar system we have, unless energy was removed from the system from without or unless the planets were formed with just the right speed at the distances from the sun they currently occupy. Of course, if we embrace the latter possibility, then gravity has done no work in explaining how order came to arise in the solar system. So it looks as if we are forced to accept that there just aren't any initial states of the chaos that would result in the formation of our solar system other than the current state of the solar system.

To this we should add Bentley's final argument from gravitation: without intervention (presumably by God) the universe as we know it is not stable. The stars, though quite distant from one another, nonetheless perpetually attract if gravitation is universal. If left to their own devices, they should all collapse to a single massive lump of matter. That they haven't done so suggests that either gravitation is not universal – in which case the providential selection of which things gravitate and which don't suggests a divine influence – or something intervenes to keep the stars from pulling one another in, which again suggests a divine influence.

5.4 Nieuwentyt

In 1716, a Dutch mathematician named Bernard Nieuwentyt published
a book on natural theology in the Low Dutch language. The book was
quickly translated into English by John Chamberlayne and appeared as *The
Religious Philosopher; or the Right Use of Contemplating the Works of the Creator.*
In section XXIX of the Preface (the English version of the book was published in three volumes and the Preface alone spans thirty-three sections),
Nieuwentyt provides a short summary of the argument developed in the
bulk of the text. The argument goes like this: nature is full of examples
of systems with some special characteristics. By a system I just mean a
collection of interacting parts where each part simply obeys the set of
physical laws – each part is utterly incapable of intentionally directing its
own action. Examples of systems of this sort include things like rivers and
piles of sand as well as things like machines and organisms. Even people
are systems in this sense; while each person might be capable of intentional action, each of her parts (e.g. cells or molecules) is presumably not.
Now machines and people are prime examples of systems with the special characteristics Nieuwentyt has in mind, namely, the parts together
accomplish some function "of great Use and Service, and sometimes of the
Utmost importance."[11] Furthermore, the repeated mechanical interaction
of the parts of the system can repeatedly or continuously produce this
function – it's not a one-time affair. What's more, in the special systems of
interest, removing a single part or altering the properties of a part just a
little destroys the ability of the system to satisfy the useful function. These
characteristics, says Nieuwentyt, are precisely those that result from intelligent agency, from intentional design, and only result from intentional
design. In other words, Nieuwentyt gives us an argument from order in
which 'order' is understood as the collection of special system characteristics mentioned above. Since these characteristics can be found in living
things as well as the technological products of human design, we must
infer that at the very least there exists a wise designer of living things.

To illustrate this rather abstract argument, Nieuwentyt provides a striking example. He asks us to imagine that "in the middle of a Sandy down,
or in a desart [*sic*] and solitary Place, where few People are used to pass,

[11] Nieuwentyt 1721, xlvi.

any one should find a Watch, shewing the Hours, Minutes, and Days of the Months."[12] Upon examination of the watch, our wanderer discovers that it contains many different toothed wheels, adapted to one another such that "one of them could not move without moving the rest of the whole Machine." Further, he notes that the wheels are made of brass and so not prone to rust. The spring which drives the toothed wheels is of steel, which is the best suited metal for making springs. The face of the device is covered by a clear glass so that the hand which is turned by the wheels can be seen without opening the case and yet remains protected. He discovers as well a key that exactly fits a pin on the back of the watch and which, when turned, winds a chain which compresses the spring, which in turn sets the whole assemblage into regular motion. Most importantly, "He would perceive, that if there were any Defect either in the Wheels, Spring, or any other Parts of the Watch; or if they had been put together after any other Manner, the whole Watch would have been entirely useless." In other words, if any of the parts were altered only slightly in form or function, the entire object of the watch would fail to exhibit its most notable behavior, namely the regular motion of its hands. Nieuwentyt suggests that this clear adaptation of parts for an end – the function of marking time – entails the agency of an intelligent being with the end of producing a device that marks time. The same reasoning could be applied to other artificial works like ships, houses, and paintings. In all cases, the conclusion we have reached follows with "mathematical certainty" as near as any conclusion does – Nieuwentyt views the argument as a deduction. Nieuwentyt then claims that all of the properties which allow us to infer design for a purpose in the watch can be found in the "Construction of the visible World." Living things possess many parts intricately adapted to one another so as to result in a useful function that is destroyed the moment one part is altered. For instance, the form and placement of molars in the mouth is just such as to permit the grinding of food. Thus, we can conclude that all such parts of the visible world – all living things, for instance – are the products of design.

The structure of Nieuwentyt's argument is shown in Figure 5.1. I won't examine its strengths and weaknesses here since we'll see the same argument again when we examine the work of William Paley in Chapter 8.

---

[12] Nieuwentyt 1721, xlvi–xlvii.

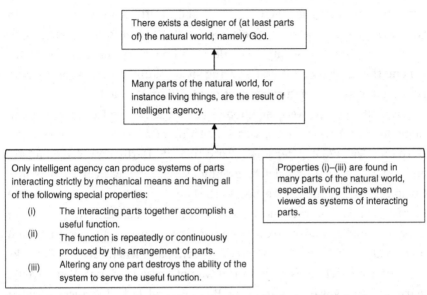

There exists a designer of (at least parts of) the natural world, namely God.

Many parts of the natural world, for instance living things, are the result of intelligent agency.

Only intelligent agency can produce systems of parts interacting strictly by mechanical means and having all of the following special properties:

(i)   The interacting parts together accomplish a useful function.

(ii) The function is repeatedly or continuously produced by this arrangement of parts.

(iii) Altering any one part destroys the ability of the system to serve the useful function.

Properties (i)–(iii) are found in many parts of the natural world, especially living things when viewed as systems of interacting parts.

Figure 5.1   Nieuwentyt's design argument, which is a kind of argument from order

It is important to note that the conclusion of the argument is explicitly intended to follow with mathematical certainty (or nearly so), and is thus a variety of the argument from order. We should therefore treat the specific properties of order that Nieuwentyt identifies in the watch as jointly sufficient conditions for inferring intelligent agency.

5.5 Ray and Derham

The arguments of Boyle, Bentley, Newton and others who drew upon the new physics were mathematically abstruse. But interest in design arguments was broad, especially given the widespread if somewhat baffling perception that Christendom was in the midst of a surge in atheistic tendencies and moral decay. This state of affairs created a demand for more accessible arguments from the burgeoning sciences. Nieuwentyt's argument – which cites details of the natural world that require no specialized mathematics to understand – was one such argument, and it enjoyed popularity no doubt in part because of its relative accessibility. In the seventeenth and early eighteenth centuries, the subjects of biology, geology, and meteorology were still largely the province of natural history, consisting of catalogues of curious facts and careful compilations of observation. Even so, these subjects offered fertile ground for producing arguments that, like Nieuwentyt's, could be followed by the general public. Among the most successful at crafting such arguments were the naturalist John Ray and the naturalist-physicist William Derham.

Ray and Derham were neighbors, close friends, and collaborators in England. Ray, born in 1628, was something of a mentor to the younger Derham (1657–1735). After Ray's death in 1705, Derham published a panegyric that included a brief biography as well as Ray's correspondence. Curiously, the biographical aspect of the book – entitled *Select Remains of the Learned John Ray*[13] – reads more like a résumé than a life history. Derham tells us when Ray undertook which major research expeditions in England or abroad, when he made various discoveries, and when each of his books was published. There is little in the way of personal detail. What we do learn is that Ray was born in Essex. By the age of 15 he had received what was possible by way of education at the local school and was sent off to Cambridge. After earning his Master's, he became a lecturer at Trinity College (of Cambridge University), where he was known for his moving sermons. He was ordained in 1660 and admitted as a Fellow of the Royal Society in 1667. While his principal expertise was in botany – he published a *Catalogue of Cambridge Plants*, a treatise on the movement of sap in trees, and accounts of his botanical observations in Europe – he was perhaps most famous for his 1691 book on natural theology: *The Wisdom of God Manifested in the Works of Creation*. The book was well received and published in some five editions.

Derham picked up where his mentor left off. Derham was educated at Trinity College in Oxford, not Cambridge. He was ordained in 1682 and was granted an ecclesiastical living at Upminster church. From the tower of this church, by timing the delay between the visible flash and audible bang of cannon being fired 12 miles away across the River Thames, Derham was able to compute the most accurate speed of sound obtained by that time. When Newton published the second edition of the *Principia* in 1713, he had adjusted his calculations to recognize Derham's result. Derham's most famous works are a pair of large treatises on cosmology and natural history that are supposed to constitute one long argument

---

[13] Derham 1760.

for the existence of God, just like Ray's *The Wisdom of God*. These remarkable compendia of results in astronomy, physics, meteorology, and natural history were entitled *Physico-Theology* (1713) and *Astro-Theology* (1715). Each book began as a series of Boyle lectures for the years 1711 and 1712 respectively.

Both Derham and Ray rely on the old Stoic arguments from order, providence, and purpose that we introduced in Chapter 3, though neither author is very explicit about it. When they are explicit in their arguments, both authors tend to jumble the three types together. Since we have already catalogued the arguments from order, providence, and purpose, I'll present just a few examples of where, like the character Balbus in Cicero's dialogue, Ray and Derham appear to make deductive arguments from the existence of order or of purpose or of providence in the parts of the cosmos. Then we can ask what advance these authors have made over their Stoic predecessors.

With respect to purpose Ray infers design as follows:[14]

> The admirable Contrivance of all and each of them [God's creations], the Adapting all the Parts of Animals to their Several Uses: The Provision that is made for their Sustenance … And, Lastly, Their mutual Subserviency to each other and unanimous conspiring to promote and carry on the Publick Good, are evident Demonstrations of His Sovereign Wisdom.

Amongst many others, Derham cites as an example of the purposeful adaptation of means to ends the gall wasps, which you might recall from Chapter 1. He speaks with evident and well-justified wonder at the ability of the tiny wasps to induce a vastly different sort of organism (mostly trees) to produce beautifully structured homes for them. He describes the many varieties of gall and the remarkable specificity with which each species of wasp causes its host tree to produce the gall characteristic of that species. His ample footnotes are full of fascinating bits of natural history, including a guess at the existence of further insects that parasitize the gall wasps by injecting their eggs in the gall. He concludes his discussion this way:[15]

> And now, these things being curiously considered, what less can be concluded, than that there is a manifest design and forecast in this case,

[14] Preface to Ray 1714.   [15] Derham 1786a, 105.

and that there must needs be some wise artist, some careful, prudent conservator, that from the very beginning of the existence of this species of animals, hath, with great dexterity and forecast, provided for its preservation and good! For what else could contrive and make such a set of curious parts, exactly fitted up for that special purpose; and withal implant in the body such peculiar impregnations, as should have such a strange uncouth power on a quite different rank of creatures? And lastly, what should make the insect aware of this its strange faculty and power, and teach it so cunningly and dexterously to employ it for its own service and good!

For Derham, the world of living things abounds in evident purpose and the adaptation of means to obvious ends.

Consider next this statement of the central argument in Ray's *The Wisdom of God*:[16]

There is no greater, at least no more palpable and convincing Argument of the Existence of a Deity than the admirable Art and Wisdom that discovers itself in the Make and Constitution, the Order and Disposition, the Ends and Uses of all the Parts and Members of this stately Fabrick of Heaven and Earth: For if in the Works of art, as for example a curious Edifice or Machine, Counsel, Design, and Direction to an End appearing in the whole frame, and in all the several pieces of it, do necessarily infer the Being and Operation of some intelligent Architect or Engineer, why shall not also in the Works of nature, that grandeur and magnificence, that excellent contrivance for Beauty, Order, Use, &c. which is observable in them, wherein they do as much transcend the Effects of humane Art as infinite Power and Wisdom exceeds finite, infer the Existence and Efficiency of an Omnipotent and All-wise Creator?

This is clearly an appeal to order – to properties that can serve as marks of design and allow us to infer with certainty the agency of a Designer. While considering astronomy and the general behavior of the cosmos, Derham makes a similar case by quoting directly from Cicero:[17]

But the perennial, and perpetual courses of those stars, together with their admirable and incredible constancy, declare a divine power and mind to be in them.

---

[16] Ray 1714, 30.
[17] Derham 1786a, 294 (*Astro-Theology*, Book IV, ch. vi).

Or more expansively:[18]

> What motion, what contrivance, what piece of clock-work, was there ever
> under the whole Heavens, that ever came up to such a perfection [as the
> motion of the Earth and Heavens], and that had not some stops or some
> deviations, and many imperfections? But yet none was ever so stupid as
> to conclude such a machine (though never so imperfect) was made by
> any other than some rational being, some artist that had skill enough for
> such work. As he in Cicero argues from his friend Posidonius's piece of
> watch-work, that shewed the motions of the Sun, Moon, and five [planets];
> that had it been carried among the Scythians or Britons ... no man even
> in that state of barbarity would make any doubt, whether it was the
> workmanship of reason or no. And is there less reason to imagine those
> motions I have been treating of to be other than the work of God, which
> are infinitely more constant and regular than those of man!

We can even find a combination of the arguments from order and purpose, suggesting that the order in the motions of celestial bodies is clearly
contrived to satisfy the purpose of providing for human life:[19]

> And now to reflect upon the whole, and so conclude what hath been said
> concerning these several motions; we may all along perceive in them such
> manifest signals of a Divine hand, that they all seem, as it were, to conspire
> in the demonstration of their infinite Creator and Orderer. For besides
> what, in all probability, is in other parts of the universe, we have a whole
> system of our own, manifestly proclaiming the workmanship of its maker.
> For we have not these vast and unwieldy masses of the Sun, and its planets,
> dropt here and there at random, and moving about the great expansum, in
> uncertain paths, and at fortuitous rates and measures, but in the completest
> manner, and according to the strictest rules of order and harmony; so as
> to answer the great ends of their creation, and the divine providence; to
> dispatch the noble offices of the several globes; to perform the great works
> of nature in them; to comfort and cherish every thing residing on them, by
> those useful changes of day and night, and the several seasons of the year.

As for the argument from providence, recall that it goes like this:

> Many aspects of the world are configured precisely as they must be for the
> continued existence, flourishing, and enjoyment of human beings.

---

[18] Derham 1786a, 290–91.   [19] Derham 1786a, 292.

Therefore, the universe was designed for the benefit of humans.

Thus, there exists a designer of the universe.

This sort of inference is presented frequently in both the works of Ray and Derham. Here's a typical example from *Physico-Theology*:[20]

> So another plain sign of the same especial providence of God, in this matter, is that generally throughout the whole world, the earth is so disposed, so ordered, so well-laid; I may say that the mid-land parts, or parts farthest from the sea, are commonly highest; which is manifest I have said, from the descent of the rivers. Now, this is an admirable provision the wise Creator hath made for the commodious passages of the rivers, and for draining the several countries, and carrying off the superfluous waters from the whole earth, which would be as great an annoyance as they now are a convenience.

In short, the topography of the continents reveals a precise jiggering of the universe for the benefit of humankind. Thus, the universe was created by a God who wishes to provide for us.

As I said, the arguments to be found in the works of Ray and Derham are old ones, stated without much difference from the words of Cicero himself. So why include them? What did the seventeenth and eighteenth centuries have to offer that was lacking in 100 BCE? The answer is examples – detailed, carefully documented examples. What Ray and Derham added to the Stoic arguments was an enormous breadth of observational and experimental data on the form and function of the natural world, particularly living things. The scope of Derham's work in particular is breathtaking; his two books survey the state of the art in astronomy, geology, meteorology, botany, zoology, human physiology, and even human demographics. One cannot help but be impressed by the range of erudition reflected in the table of contents of his books, especially once one realizes the extent to which Derham contributes to the subjects he surveys. On top of adding an enormous number of detailed premises where the Stoics drew on a handful of vague examples, the works of Ray and Derham and their imitators were, as I suggested above, much more accessible to an interested lay audience than the arguments of Bentley or Newton. As Ray notes in the Preface to *The Wisdom of God*, an advantage of his arguments is that,

---

[20] Derham 1786b, 108.

while capable of persuading the learned, they are "intelligible also to the meanest capacities." This sentiment is echoed by the Scottish mathematician Colin Maclaurin. He wrote that the arguments from the apparent adaptedness of organisms and other parts of the cosmos for one another are "sufficiently open to the views and capacities of the unlearned, while at the same time they acquire new strength and luster from the discoveries of the learned."[21] That about sums up the project of Ray's *The Wisdom of God* and Derham's *Physico-* and *Astro-Theology* – adding strength and luster to old and simple arguments.

---

[21] Maclaurin 1750, 400–1.

6 Unusual design arguments

6.1 Overview

The arguments of the last chapter were recycled through seemingly end-less variations during the two hundred years I've been calling the golden age of natural theology. As with Ray and Derham, the Stoic arguments from order and providence provided a sort of loose template that many a naturalist and theologian filled with examples from the natural world. However, the golden age did produce a few arguments of a decidedly dif-ferent flavor. In this chapter, we explore attempts to demonstrate the existence of God from our experience of a 'Visual Language', of beauty in the world, or from our direct perception of design.

6.2 A bit about Berkeley and the *Alciphron*

The argument from the existence of a Visual Language comes to us from George Berkeley,[1] Bishop of Cloyne, by way of his once famous and strangely titled book, *Alciphron, or the Minute Philosopher*. Berkeley was noth-ing if not original. Born in Ireland in 1685 – the same year as J. S. Bach and G. F. Handel – Berkeley was a cleric and a philosopher of the first rank. He was a generation behind Hobbes and Locke, a contemporary of Newton, Pope, and Hume. Berkeley's philosophy was a sort of idealism which fol-lows from taking the program of the British empiricists – particularly Locke – to its extreme.

Very roughly, Locke, like the Stoics, was a concept empiricist. That is, he thought that all knowledge consists in the recognition of relations between

---

[1] Berkeley's name is pronounced 'Bark-lee'. Despite the difference in pronunciation, the city of Berkeley, California is in fact named for the philosopher.

concepts such as similarity or contiguity in time or space. Concepts in turn were derived strictly from experience, whether of the outside world via the senses or of our own mental life via introspection. We are all born *tabulae rasae* ("blank slates") according to Locke; there are no innate concepts. Of course, in order for knowledge to bear any relation to an external world of objects, there must be a close relationship between some or all concepts and real, mind-independent things like desks, laptops, and classrooms. Locke thought that concepts (or 'ideas' in his terminology) were caused by the properties (or 'qualities') of objects. Qualities, according to Locke, are of two kinds: primary and secondary.[2] The primary qualities of objects are special in that the ideas they cause in us exactly resemble the objects themselves in some way. These qualities include shape, number, and motion. Secondary qualities, like color, are ideas likewise caused by the interaction of objects with our organs of sense but which in no way resemble the object itself. This distinction accords rather well with the modern understanding of material structure. Volume is a property of an object like a coffee mug, which is made of atoms bound together. But color is not a property of any of the atoms, only a consequence of the way they scatter light and the way our eyes and brains translate wavelengths of light into experiences. At any rate, if we accept Locke's account, knowledge involving concepts caused by primary qualities is really knowledge of the external, mind-independent world.

In his most important works,[3] Berkeley attacked Locke and the other empiricists for violating their own principled appeal to experience. He pointed out – particularly in A *Treatise Concerning the Principles of Human Knowledge* and in *Three Dialogues between Hylas and Philonous* – that there is nothing in experience on which to base the distinction between primary and secondary qualities. That is, there is nothing inherent to our ideas that would allow us to tell which are exact copies of the real properties of objects and which are not – they are all just ideas. For instance, I perceive color every bit as vividly as shape and there is nothing to distinguish the two in my experience such

---

[2] He actually distinguished three sorts of qualities: primary, 'secondary immediately perceivable', and 'secondary mediately perceived'. The distinction between the latter two is not important for us here, but for Locke's own explication, see Book II, ch. viii of the *Essay Concerning Human Understanding*. See especially section 26 of that chapter for the terminology cited here.

[3] See Copleston 1993, Vol. V, ch. 11, section 2, 203–4.

that I can say shape is truly a property of things outside my mind while color is not. In fact, Berkeley used the very same arguments Locke had introduced to show that ideas like color and odor cannot correspond to properties in objects in order to argue that the ideas corresponding to Locke's primary qualities can't be *in* objects either. Without the distinction between ideas caused by primary and secondary qualities, there is nothing to tie our knowledge to the external world – if experience does not give us a way to tell which ideas are copies of real properties and which aren't, then experience does not provide any reason to believe that there exist mind-independent objects at all. Thus, we arrive at Berkeley's idealism – the doctrine that only ideas exist. It is concept empiricism in its purest form. This is the philosophical stance lurking under the surface of *Alciphron*.

When Berkeley wrote *Alciphron*, he was actually in America. He had sailed to Newport, Rhode Island with the intent of establishing a missionary college in Bermuda, having been granted a charter and promised financial support by Lord Walpole (effectively the first British prime minister). As Berkeley saw it, the missionary college would be "for promoting reformation of manners amongst the English in our Western plantations, and for the propagation of the Gospel amongst the American savages."[4] As Copleston puts it, "[h]e apparently envisaged English youths and Indians coming a very considerable distance from the mainland of America for general, and especially religious, education, after which they would return to the mainland."[5] Though he later revised his plans and petitioned parliament to construct the college in Rhode Island where he'd been trying to manage the logistics, the funding never materialized and he returned to England. In the three years he'd spent in Newport – from 1728 to 1731 – he managed to write *Alciphron, or the Minute Philosopher*, which was published upon his return to England.[6]

Alciphron came late in Berkeley's career. As the effusive Preface to the first American edition puts it, "[t]he Minute Philosopher consists of a series of dialogues, involving most of the important topics in the debate between Christians and Infidels; the principal arguments by which Christianity is defended, and the principal objections with which it has been opposed."[7] The so-called 'freethinkers' – those skeptical of revealed religion – are

---

[4] These are Berkeley's words from a letter to Lord Percival, quoted by Fraser in Berkeley 1901, 5.

[5] Copleston 1993, Vol. V, 202.   [6] Copleston 1993, Vol. V.   [7] Berkeley 1803.

represented principally by the character Alciphron, "or the 'strong man' in his own conceit."[8] The principal theme of the dialogues is the supposed narrowness or smallness of the intellectual horizon of the freethinkers, hence the phrase 'minute [tiny] philosophy'. It is Euphranor who establishes the rationality of religious belief and, in the dialogue we read, the very existence of God, whilst Crito largely moderates. The remaining characters play mostly a literary role. Lysicles is the friend of Alciphron and a fellow freethinker whose character provides further opportunity for Berkeley to mock his foe. Dion is, according to Fraser at least, Berkeley's avatar in the dialogue, and remains largely silent.[9]

The first dialogue[10] sets the ground rules for the rest of the debate, with Alciphron conceding that the proper measure of the morality of an act or institution is its tendency to promote the good of mankind. Thus, the central question of this Christian apologetic becomes: does belief in God, heaven, and so forth tend to promote the good of mankind? The second dialogue is intended as a refutation of the thesis of Mandeville's satirical *Fable of the Bees*, namely that the vices of individuals which make each of them happy thus tend to promote the happiness of society and, in accord with the premise above, must be viewed as universally moral. Berkeley responds that the social ills which accompany individual vices outweigh their contributions to happiness, and that (à la Mill) there are different kinds of pleasures to worry about. In the third dialogue, Berkeley rebuts a thesis which he attributes to Shaftsbury that conscience can be reduced to taste – the morality of an act is conventional. In the fourth dialogue, Berkeley takes up what is really the central question to be settled before the value of religious belief can be weighed: is there any *reason* to believe in a God? We'll look at his argument in some detail in the next section, but it suffices to note here that he thinks he has a knock-down argument for the existence of God consistent with empiricism. The last three dialogues of the book comprise "a vindication of religion in its Christian form."[11] That is, after establishing God's existence and a criterion for moral valuation, Berkeley goes on to defend the details of Christian dogma and ritual.

[8] Berkeley 1901, B3.   [9] Berkeley 1901, 9.

[10] With the exception of the fourth dialogue, the synopses presented here draw upon Fraser's discussion in Berkeley 1901, 9–11.

[11] Berkeley 1901, 11.

6.3 The Visual Language argument

Berkeley's central argument in *Alciphron* is a design argument, one that borrows an important idea from René Descartes that has nothing directly to do with God. Instead, it has to do with the existence of other minds. Here's the puzzle: you know you have a mind. But how do you know that the rest of us with human bodies also have a mind like yours? How do you know we're not just really sophisticated wind-up toys or zombies? Descartes famously claimed that no matter how lifelike a human-shaped automaton one can make, it could always and immediately be distinguished from a real person using a simple pair of tests. Here's how Descartes puts it in Part V of his *Discourse on Method*:[12]

> [I]f any such machines resembled us in body and imitated our actions insofar as this was practically possible, we should still have two very certain means of recognizing that they were not, for all that, real human beings. The first is that they would never be able to use words or other signs by composing them as we do to declare our thoughts to others. For we can well conceive of a machine made in such a way that it emits words, and even utters them about bodily actions which bring about some corresponding change in its organs (if, for example, we touch it on a given spot, it will ask what we want of it; or if we touch it somewhere else, it will cry out that we are hurting it, and so on); but it is not conceivable that it should put these words in different orders to correspond to the meaning of things said in its presence, as even the most dull-witted of men can do. And the second means is that, although such machines might do many things as well or even better than any of us, they would inevitably fail to do some others, by which we would discover that they did not act consciously, but only because their organs were disposed in a certain way. For, whereas reason is a universal instrument which can operate in all sorts of situations, their organs have to have a particular disposition for each particular action, from which it follows that it is practically impossible for there to be enough different organs in a machine to cause it to act in all of life's occurrences in the same way that our reason causes us to act.

In this single paragraph, Descartes offers us two ways to detect other minds. The first test – the one Berkeley uses – rests on the premise that only minds are capable of using *language*. Descartes doesn't mean simply that we can make noises or other signs in response to stimuli. Rather, genuine language involves a special use of symbols. In particular, languages are 'generative' in that you can generate an endless variety of expressions from a relative handful of words or sounds. With a modest English vocabulary and the rules of English grammar (such as they are) one can produce an endless variety of sentences. Very few other animals use signs – verbal or otherwise – that can be stitched together like our words to make new, meaningful messages. Those that do – chickadees for instance – have a very limited repertoire (not that Descartes was aware that chickadees have grammar). Furthermore, language use is responsive – we don't just build sentences, we build sentences that say useful and relevant things about our states of affairs, our goals, what we want others to do, etc. A dead giveaway for a machine, says Descartes, is its inflexibility. Consider, for instance, the obnoxious phenomenon of the robo-operator. When you call any large company these days, you end up speaking to a computer, not a person. The computer will utter some prerecorded greeting, then give you a menu of options: "For sales, press or say 'one', for technical support, press or say 'two' …" It responds to what you then say in only very limited ways, and in every case can only give you a prerecorded response. It doesn't stitch together new combinations of words to address you in the context of the conversation. It never says, "I'm terribly sorry to hear that your laptop exploded. Is everyone all right?" And getting to the right department is hopeless if you give an answer that is outside its short list of recognized words. Of course, as I describe the example I am aware that researchers in artificial intelligence and human–computer interaction are working to change this state of affairs – we'll get to that below. The point is just that we can tell a fake, a machine or zombie with no mind, because it fails to use its putative language adaptively. What Descartes takes as a separate test – namely that real minds are good at a huge range of tasks – goes to emphasize this aspect of the first test. Genuine language use is adaptive in a wide range of situations. Only a mind can use language in a way that is responsive to all conceivable circumstances.

Berkeley realized that we can treat the question of design in the world as a special problem of other minds. How could I know whether or not the natural world – or parts of it – are associated with a mind? Administer the language test! This is Berkeley's strategy in the *Alciphron*. But before we can use it, we have to clarify a very important point. Descartes mentions "words or signs," but doesn't tell us what it takes for a sound or a written character to count as a word or sign. Before we can have an adaptive language based on a grammar, we have to at least have some signs or symbols. What does it take for something to be a symbol? Berkeley says that a symbol is something like a vocalization or written character that corresponds to an event, object, or state of affairs in an *arbitrary* way. Lots of things are correlated with other things, but don't count as symbols. For instance, lightning tends to precede thunder, but lightning is not a symbol for thunder. The connection between a symbol and what it signifies is special in that there is no logical or physical reason that compels that particular association. Think of the words you're reading on this page. The collection of letters 'cat' is a symbol that refers to or signifies a particular sort of animal. But there is no reason we had to choose that combination of letters to stand for that animal. We might just have well have used 'cat' to signify a hippopotamus, or clouds, or anything else. Likewise, we could have associated the word 'geederplex' with the type of animal currently signified by 'cat'. The point is that language uses symbols, and symbols are only arbitrarily connected to that which they signify. That's why we don't think thunderstorms are using language.

A. David Kline distills Descartes' language test (as it is clarified by Berkeley) into three criteria that must be met for a collection of signs to count as a language:[13]

(1) It is composed of signs or symbols which bear no necessary connection to the thing signified – the association between symbols and a language and that which they symbolize is arbitrary.
(2) It is *generative*, meaning that it has a grammar according to which signs can be meaningfully combined in a diverse (or even unlimited) number of ways.
(3) It is appropriate and responsive.

If all three criteria are met, then we have an instance of genuine language use and therefore, according to both Descartes and Berkeley, certain evidence of another mind.

In his dialogue, Berkeley has the "minute philosopher," Alciphron (the atheist), suggest the Cartesian language test and explain the criteria. He

---

[13] Kline 1993.

asserts that we can infer with great confidence that other people possess minds because we witness their use of language. Here's how Alciphron puts it:[14]

> What I mean is not the sound of speech merely as such, but the arbitrary use of sensible signs, which have no similitude or necessary connexion with the things signified; so as by the apposite management of them to suggest and exhibit to my mind an endless variety of things, differing in nature, time, and place; thereby informing me, entertaining me, and directing me how to act, not only with regard to things near and present, but also with regard to things distant and future. No matter whether these signs are pronounced or written; whether they enter by the eye or ear: they have the same use, and are equally proofs of an intelligent, thinking, designing cause.

Berkeley, through his mouthpiece Euphranor, endorses the test as sound and promises to use it to extraordinary effect. Alciphron the atheist is compelled to accept the outcome come what may:[15]

> *Euphranor*: But if it shall appear plainly that God speaks to men by the intervention and use of arbitrary, outward, sensible signs, having no resemblance or necessary connexion with the things they stand for and suggest: if it shall appear that, by innumerable combinations of these signs, an endless variety of things is discovered and made known to us; and that we are thereby instructed or informed in their different natures; that we are taught and admonished what to shun, and what to pursue; and are directed how to regulate our motions, and how to act with respect to things distant from us, as well in time as place, will this content you?
> *Alciphron*: It is the very thing I would have you make out; for therein consists the force, and use, and nature of language.

Berkeley's strategy, then, is to take Descartes' argument for other minds as sound, and to apply it in a new way in order to infer the existence of God. Rather surprisingly, he argues that visual sensations – the elements of our visual experience of the world – meet the criteria for being a language when treated as a collection of messages. We can then conclude that the greater part of our everyday experience is therefore evidence of the existence of a mind constantly speaking to us through our sense of sight.

---

[14] Berkeley 1901, 163.　　[15] Berkeley 1901, 163–64.

It is not what experience tells us about the world but experience itself that forms the key premise. Now that's an unusual design argument.

Of course, Berkeley first has to convince us that visual sensations meet the criteria for a genuine language. Let's take each of these in turn. First, he argues that particular visual sensations are arbitrarily connected to tactile experiences. In other words, he argues that visual sensations are informative *symbols* for tactile objects like trees, other people, pet cats, laptops, distant buildings, etc. Importantly, he has to show that they are informative not because they are copies of the original, but rather are arbitrarily connected to the things they inform us about. To convince us that our visual sensations are symbols, Berkeley needs to convince us that what we see is not connected to what it informs us about in a way that is logically or physically necessary – it could have been different. You might think this is hopeless because what you see are just objects. I look in the corner and see a chair; I don't see a symbol. But here is where Berkeley marshals some of the arguments that underlie his idealism. To use one of Berkeley's own examples, step outside and look up at the clouds. From the ground on a sunny day, cumulus clouds – the puffy ones that look like cotton balls – appear white and opaque. However, whiteness and opaqueness are not properties of the actual cloud. If we went to the cloud, in a balloon say, it would appear as a fine, translucent mist. So whatever properties the object itself has, these are not identical to the properties of your visual sensations. To put it another way, the objects of vision – the sensations with color, intensity, etc. – are not identical with the objects themselves. Of course, for Berkeley the conclusion is that there are no objects themselves, only collections of sensations. But we needn't go that far. It's enough that what you see is not the object but an image associated with an object.

Furthermore, features of your visual sensations are *arbitrarily* connected to other properties of objects, in particular the tactile sensations they produce in us. To update another example of Berkeley's, imagine you're riding in a car on a highway. Looking out the windshield, you have the visual sensation of a very small, green rectangle. You know right away that you are looking at a highway sign that is far away. As the car moves on, the image of the sign grows in size and sharpness – you know from this that you are getting closer. Berkeley points out that things might have been different. It does not follow as a logical consequence of having vision that distant things must appear small. It might have been such that distant things appear large. Or purple, or fuzzy and yellowish. The point is, any association between properties of visual sensations and properties of the object is at least logically possible. It also appears that the associations we have are not physically necessary either (more on this below). So the connection between visual properties and objects or object properties is symbolic.

We have next to show that the visual language is generative. In other words, that there is an unlimited variety of ways to combine visual cues into meaningful visual messages. This step is rather easy – just look around. If you accept that visual sensations are symbols, it is immediately obvious from experience that there seem to be a more or less boundless number of ways in which these sensations can be combined. Color and intensity can vary continuously across your whole visual field. To put it in modern terms, you can imagine that visual sensations are built up out of pixels, each pixel with a particular intensity of red, green, and blue. Every conceivable combination of pixel values is a viable visual image or, in this case, symbolic message.

But is it really the sort of message that can only come from an intelligent source? The last criterion in our three-part test is that the messages generated by a putative speaker of the language (in this case, the source or author of our visual sensations) must be adaptive. As conditions change in the world, our visual sensations adjust and remain informative. It's not enough that we receive a variety of messages. To know that they originate with a mind, the messages must be adaptive, appropriate to our circumstances, no matter how those change. Here again, says Berkeley, you need only consider a sliver of your everyday experience of the world. Suppose your visual sense was like the robo-receptionist, spitting out the same answer no matter how circumstances change. Then you would continue to see the same thing in cases in which objects actually move or shift properties. But this is not the case. As you move about the world and objects interact with one another, your visual sensations keep pace – you receive a steady stream of fresh messages that continue to accurately inform you about the world. So it seems that our visual sensations do in fact pass the language test. If so, then we can infer the existence of a mind that is using

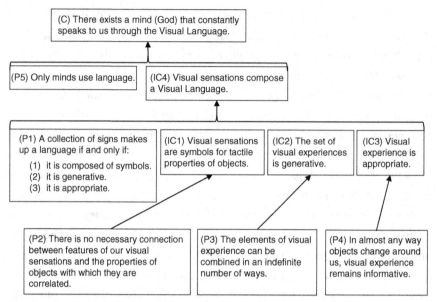

Figure 6.1   Berkeley's 'Visual Language' argument

that language. That is, someone is speaking to us through our visual sense. A mind that powerful surely warrants the title 'God'. Here is the outcome of the analysis summarized in Berkeley's own words:[16]

> Infinitely various are the modifications of light and sound, whence they are capable of supplying an endless variety of signs, and, accordingly, have been much employed to form languages; the one by the arbitrary appointment of mankind, the other by that of God Himself ... But such as the connexion is of the various tones and articulations of voice with their several meanings, the same is it between the various modes of light and their respective correlates, or, in other words between the ideas of sight and touch.

This is a very unusual argument, and to see how all the pieces are supposed to fit together, it might help to consider the diagram in Figure 6.1.

6.4 Assessing the Visual Language argument

The question for us is whether the argument is sound or cogent. As I've presented it in Figure 6.1, the argument is deductive. But we might reasonably

---

[16] Fraser 1911, 275–76.

take Descartes' test to yield only a probable conclusion. There are a couple of ways to do so. The criteria for language in (P1) might be considered only a probabilistic test: if a collection of signs meets the three criteria it is probably a genuine language. Or we might weaken (P5) and accept that some things without minds can genuinely be said to use a language, but not many things. Thus, language use makes it only probable that something has a mind. However we construe the argument – deductive or inductive – most significant criticisms center on premise (P5). To explore a couple of these, let's take the characterization of language in (P1) as a definition, and consider how plausible (P5) is. Unlike Descartes or Berkeley, we live in an age in which machines win against human contestants on game shows like *Jeopardy!* that require verbal acumen, our cars can tell us how to get from one point to the next and dynamically update their instructions as the driving situation changes, and we can ask questions of our 'smartphones' in plain English and get more or less sensible replies. It is tempting to think that perhaps machines do use language. Perhaps it is more plausible to consider language use on a continuum – it's not something that is either there or not, but comes in degrees. Some animals have modest grammars and use language to a modest degree. People are highly skilled users of language. In that case, our best artificial intelligence algorithms fall somewhere in between, using language to an intermediate degree. Of course, this is only an objection to (P5) if we insist that machines that use language have no minds. This is a somewhat controversial claim. One might be tempted to agree with Descartes that full-blown language use of the sort exhibited by humans really is indicative of mind, and so we should treat any machine capable of that kind of linguistic feat as actually possessing a mind. Whether or not modern technology provides an objection to (P5) depends on what we take minds to be and what position we take on what is necessary to instantiate a mind.

Let's consider one more retort to Berkeley involving (P5). As a modern reader, you might wish to tell the following story to Berkeley. We evolved in an orderly world. That is, we are organisms who live in a world governed by physical laws. Our sensory systems and the mental sensations they induce were developed over a great many generations. Those organisms with the ability to detect and represent environmental conditions with their sense organs and brains were favored and survived. Over time, these sense organs and neurological representations got really good. The sensations Berkeley speaks of are just the mental representations of our sensory states. The point is that it is no surprise that these sensations do a good job of representing the world – they were evolved to do so!

We haven't yet discussed the biological theory of evolution and its ramifications for design arguments. The essence of the objection, however, is that there are conceivable processes that do not involve a super-mind feeding us information whereby we could develop a 'Visual Language' with all the properties Descartes and Berkeley demand. The putative language would then be adaptive and responsive to the world not because of the mental activity of some divine speaker, but because of a long causal process of interaction between the ordered world and the human mind. In short, we are denying (P5) – other things besides minds can result in a series of signs meeting the requirements of a language laid down in (P1). Of course, for Berkeley, this is a non-starter. He thinks he has arguments that make talk of material objects and their interactions misguided – only ideas and minds exist. But he does seem to anticipate the objection, saying vaguely that the connection between visual sensations and states of the world "cannot be accounted for by mechanical principles, by atoms, attractions, or effluvia."[17] To be fair, Berkeley cannot be expected to have foreseen the sort of explanation in terms of "mechanical principles" that would be offered more than a century later. But in our assessment of the argument, this is a crucial point. If it turns out that the properties listed in (P1) can be produced without a mind, then (P5) is false and the argument fails.

6.5 Reid and Whewell on direct perception

Perhaps the best way to introduce the argument from direct perception is to think back to the example of the parasitic wasp, *Megarhyssa*, of Chapter 1. There we considered the intricate natural drill – the ovipositor – with which this insect bores through the hardened wood of a tree trunk with enough precision to inject an egg into tiny larva within. I suggested that most of us, upon hearing of this extraordinary adaptation and the details of the mechanism by which the ovipositor operates, are prone to a strong intuition of design – surely such a device was contrived by someone to serve the obvious purpose of drilling through wood and laying eggs. But

---

[17] Berkeley 1901, 174.

what if this impression of design is more than mere intuition? What if, in fact, it is perception, much like seeing that something is blue or round? That is precisely what Thomas Reid and William Whewell would have us believe.[18]

Thomas Reid was a Scottish philosopher whose life spanned nearly the entire eighteenth century (1710–96).[19] He was a contemporary and, for a while at least, follower of Berkeley. It was David Hume (whom we'll consider in the next chapter) who inspired Reid to break with Berkeley and develop his "common-sense" philosophy of knowledge and perception. Reid's theory of knowledge is foundational; he believes all justified reasoning begins with what he calls "first principles." These are[20]

> propositions that are no sooner understood than they are believed. The judgment follows the apprehension of them necessarily, and both are equally the work of nature, and the result of our original powers. There is no searching for evidence, no weighing of arguments; the proposition is not deduced or inferred from another; it has the light of truth in itself, and has no occasion to borrow it from another.

One of Reid's first principles is that there exist other human minds besides one's own. How do we know that this is a first principle and not deducible from other such axioms? Because we acquire this belief very early in life, prior to any education:[21] "the belief we have, that the persons about us are living and intelligent beings, is a belief for which perhaps we can give some reason when we are able to reason but we had this belief before we could reason, and before we could learn it by instruction. It seems therefore to be an immediate effect of our constitution." In other words, the belief in other minds is an inevitable outcome of the way we're put together – of what sort of beings we are – and therefore needs no proof. That doesn't mean we can't give arguments to support the proposition that other minds exist. But unlike the certainty that comes from the first principle, any such argument will at best give less than the perfect certainty. In fact, Reid gives us just such an argument based on another proposition

---

[18] For a modern defense of the direct perception of design, see Ratzsch 2003.

[19] Despite his longevity, Reid's biography is rather bland. For more details, the Introduction to Reid 1975 is a helpful source.

[20] Reid 1785, 555. See also the rest of Essay VI, ch. iv, for an overview of Reid's notion of first principles.

[21] Reid 1785, 574.

he claims is a first principle, namely that "from marks of intelligence and wisdom in effects, a wise and intelligent cause may be inferred." This, says Reid, is a principle whose effects we witness all the time when we 'perceive' which particular beings have minds. If I see you doing brave or wise things, I immediately and without conscious effort infer that you are a thinking thing like me and that you are brave and wise. Of course, this sounds like an inference, not a direct perception. But Reid speaks as if a conclusion that follows deductively from some perception together with a first principle may itself be considered a perception. So he would say that I 'perceive' your bravery, your wisdom, and the existence of your mind whenever I witness you engaging in acts of the right variety.

Reid thinks that we can adapt this argument for other human minds to argue for the existence of a divine mind. Here is how the design argument goes:[22]

> (1) [A]n *intelligent* first cause may be inferred from marks of wisdom in the effects. (2) There are clear marks of wisdom and design on the works of Nature – The conclusion is then – the works of Nature are effects of a designing and wise cause.

Premise (1) is the supposedly indubitable first principle of design. Premise (2) is justified by appeal to our experience of the many ways in which the universe is ideally suited to the sustenance of humankind. Naturally, the "designing and wise cause" of the conclusion is to be identified with God.

A similar argument was made by the Englishman William Whewell, a nineteenth-century scientist and philosopher of pervasive and lasting influence. In fact, it was Whewell who coined the word "scientist." Like so many of the scientists of the time, Whewell was both a member of the Royal Society and an ordained priest. Aside from prominent works in the history and philosophy of science, Whewell was invited to publish a volume in the "Bridgewater Treatises on the Power, Wisdom, and Goodness of God as Manifested in the Creation." Much as Boyle endowed a series of lectures, the Earl of Bridgewater bequeathed a sizable sum of money with instructions to the Royal Society to disburse the sum in order:[23]

> to write, print, and publish one thousand copies of a work On the Power, Wisdom, and Goodness of God, as manifested in the Creation;

[22] Reid 1981, 54; emphasis in original.
[23] Whewell 1833, ix.

illustrating such work by all reasonable arguments, as for instance the variety and formation of God's creatures in the animal, vegetable, and mineral kingdoms; the effect of digestion, and thereby of conversion; the construction of the hand of man, and an infinite variety of other arguments; as also by discoveries ancient and modern, in arts, sciences, and the whole extent of literature.

The Royal Society took its charge very seriously, and commissioned eight books from eight authors widely acknowledged to be leading authorities in their respective fields. In his volume, *Astronomy and General Physics Considered with Reference to Natural Theology*, Whewell followed a strategy similar to Reid. He argued that we cannot help but leap to the conclusion that other humans have minds, and that the same circumstances that provoke this inference are to be found in the works of nature:[24]

> No doubt their actions, their words induce us to do this. We see that the manifestations which we observe must be so understood, and no otherwise. We feel that such actions, such events must be connected by consciousness and personality ... But this is not a result of reasoning: we do not infer this from any similar case which we have known ... In arriving at such knowledge, we are aided only by our own consciousness of what thought, purpose, will are: and possessing this regulative principle, we so decipher and interpret the complex appearances which surround us, that we receive irresistibly the persuasion of the existence of other men, with thought and will and purpose like our own. And just in the same manner, when we examine attentively the adjustment of the parts of the human frame to each other and to the elements, the relation of the properties of the earth to those of its inhabitants, or of the physical to the moral nature of man, the thought must arise and cling to our perceptions ... that this system, everywhere so full of wonderful combinations, suited to the preservation, and well-being of living creatures, is also the expression of the intention, wisdom, and goodness of a personal creator and governor.

It's important to note that some of the similarity between Whewell's and Reid's argument is merely superficial. Whewell's view of scientific reasoning is rather more complicated than Reid's and more than we can reasonably treat here in any detail.[25] Whewell, who was influenced by Immanuel

---

[24] Whewell 1833, 345–46.
[25] Though for a concise overview and a brief biography, see Snyder 2009.

Kant, thought that empirical science involved a process of refining a collection of vague, innate concepts in order to arrive at necessary truths about the world. So, for instance, physics is the unfolding of the necessary consequences of the properly clarified ideas of cause in general and force in particular. For Whewell, the list of fundamental concepts is not static, but grows through time. Both the clarification of vague concepts we already have and the introduction of new ones occurs in an inductive process that depends on a special sort of inference. The inference involves adopting a hypothesis that accounts for all of the observed phenomena – that unites the phenomena under a single concept. Whewell insists that the hypotheses are not chosen arbitrarily, nor are they the outcome of induction on observed facts. However it is we come by them, we know we've gotten it right when a hypothesis accounts for all of the previously observed facts, more data of the same kind, and most importantly, data of an altogether different kind. So, for instance, we know we correctly apprehended a truth about nature – that all bodies attract one another with a force inversely proportional to the square of the distance between them – because this idea accounts for all the previously known facts of planetary motion. It also correctly predicts new facts about the motions of the planets. Furthermore, it correctly predicts and explains facts of an entirely different kind, namely the existence and nature of tides on Earth.

According to Whewell, the concept of a world-designer is a very fundamental idea – like cause or space – that accounts for a great swath of experience and can be seen to be obviously true once we stumble across it. The idea explains all of the facts about the order and adaptedness of the world in such a way that we can know it to be true. This should remind you of what I said at the very beginning of this book – one of the ways in which human beings have historically comprehended their experience of the world is in terms of minds and intention. Whewell is merely making the notion explicit and a little more precise.

What are we to make of these arguments? Let's start with Reid's argument. To begin with, there is a deep ambiguity in the claim that from marks of design we infer a designer. Reid insists that this first principle is a necessary truth. In other words, the world could not have been such that it is false. Thus it is tempting to read the principle as the innocuous and vacuous claim that instances of design in the sense of intent require an intending mind. Of course they do – it just follows from the definition of 'intent'. But that makes the principle about as informative as saying that all bachelors are unmarried. If we read the principle this way, then all the work is being done by the first premise, and Reid's argument reduces to the old Stoic argument. This interpretation is not that implausible given Reid's invocation of that very argument from Cicero (Reid 1981, 52). But then it inherits all of the same defects. Chief among these is the fact that there is nothing about a state of affairs in isolation that constitutes intent. Intent is a relation between a state of affairs and a mind. Any state of affairs could have been intended or not, there is no property of 'intended' that attaches to things by themselves. In other words, we can't tell whether something was done with purpose unless we also know what the purpose was. Of course, in the case of people, we can make good guesses as to what sorts of motives or purposes they might have.

This suggests another way of accounting for the involuntary inferences Reid attributes to a first principle of design. Think of the example of bravery. You never see bravely directly. Nor is what you see someone acting bravely – you just see someone doing things. You might, for instance, see someone running into what you know is a hazardous situation – a burning building, say – and you think (perhaps unconsciously), "Wow. If I had to run into that burning building for some reason, I would be terrified. That person is probably like me, and so he must be terrified about running into the building. Nevertheless, there he goes, clearly putting his fear aside to accomplish whatever important purpose he has in mind. He must be very brave." There is nothing wrong with this reasoning, but it is only inductive – there is no certainty in the conclusion. For instance, if what you don't know is that the person you saw is deathly afraid of spiders and is running obliviously into a burning building to get away from a particularly nasty looking spider, then the same set of actions you witnessed would fail to be brave – the person was acting heedlessly and out of panic, not bravery. You would be wrong in your inference. Now, we could suppose that you simply failed to pay attention to the right thing. But a similar argument can be made for any feature of the situation that might lead you to believe that the person is brave. So there seems to be something wrong with premise (1). It's clearly not a necessary truth. Rather, it seems to be an inductive inference, albeit one we make all the time in the case of people.

Perhaps the reason some people 'perceive' design in the works of nature is because they are illicitly applying an inductive inference that is strong when used to infer mental characteristics from the actions of people to a new case in which there is no person involved. What I'm suggesting is that when I look at various "works of Nature" and perceive design, I might really be making an inference that begins with a simple realization: it would be really hard for me to craft a device that does what this thing does. For instance, it's unlikely that it would have occurred to me to build an ovipositor of the sort *Megarhyssa* has, and even if it did occur to me the necessary materials are beyond my ability to produce. I cannot help but conclude that if a *human being like me* had produced this thing, she must be very wise indeed. Recall that for Reid, such inferences are unavoidable and unconscious – I cannot help but leap to the conclusion. But I can mistake the question to which the conclusion belongs. As I've just given it, the syllogism resting on Reid's first principle of design now reads as follows:

(1) Nature is full of states of affairs with the following property: if we were to suppose someone intended to bring about that particular state of affairs and succeeded in doing so, then that person must possess great wisdom.

(2) Therefore, there exists a wise designer of things.

This is an enthymeme – a syllogism with one of its premises left implicit. The missing premise in this case is obviously the claim that in fact someone wanted to bring about the particular state of affairs in question. But that profoundly begs the question. The suppressed premise is exactly the claim we are supposed to be proving. So if the alternative view of what's really going on when we 'perceive' design in nature is correct, then this perception fails to establish anything about the existence of a divine designer.

Of course, my alternative account of why we seem to perceive design may be entirely false. Perhaps we do look at the natural world and immediately infer from what we perceive that parts of it were designed. But there is still an inconvenient fact in the way of this view – people disagree on whether or not they perceive design. In other words, you might look at *Megarhyssa* and ineluctably conclude it was designed, while I might look and fail to see design. Assuming that both of our perceptual systems are functioning correctly, it must be that we disagree on the supposed first principle of design. That is, we must disagree on whether or not certain properties of things are marks of design that suggest a designer. But first principles are supposed to be necessary truths – it's not a matter of opinion or evidence. How are we supposed to figure out which of us is wrong? Reid concedes that such disagreement does in fact occur in the case of design, but asserts that no one familiar with natural science doubts it. That's a bit unfair. There are certainly modern scientists who profess to see no design. Surely Reid must be able to tell us *why* anyone familiar with science should accept the claim as a first principle; he must give us a justified method by which to determine who is right. Reid does claim that nature has given us the means to sort out such conflicts of intuition and to determine with close to certainty whether or not a proposition is in fact a first principle. According to Reid, absurdities should follow from either your beliefs or mine if we disagree on a putative first principle. To make Reid's argument from direct perception work in a way that doesn't beg the question, one would have to demonstrate how denial of design leads to contradiction. Reid does not do so.

As for Whewell, the argument is difficult to evaluate outside the context of his philosophical system. But even within his system, it is not obvious that we possess the right sort of evidence to secure our conviction in the hypothesis of design. We are only certain that our hypothesis has grabbed on to reality when it predicts facts of a kind different from those on the basis of which we form the hypothesis. While it is true that the hypothesis of design, when supplemented by some additional assumptions concern-ing the motives of the designer, makes sense of an enormous range of experience, it is not clear what sort of predictive value it has. What is the analog of gravity explaining the tides to which Whewell can point in favor of design? If such a case can be made, then Whewell's argument may be compelling.

6.6 Hutcheson on beauty

Francis Hutcheson (1694–1746), a Scottish philosopher often asserted to be the father of modern aesthetics, offered a lovely little argument based not on our direct perception of design, but of beauty. The argument begins by noting that we human beings have a sense of beauty. That is, certain arrangements of things such as those exhibiting symmetry and regularity elicit in us a generally pleasant but involuntary sensation we refer to as experiencing something beautiful. It is not a logical necessity that we find the particular sorts of things beautiful that we do. It is conceivable that what we now find hideous or displeasing we might instead find beautiful and vice versa. There are thus an unlimited number of ways our aesthetic tastes might have been. It is also the case that, whatever our tastes happen to be, there are an unlimited number of ways in which the world around us might be configured. But then, says Hutcheson (1726, 48),

> as there are an Infinity of *Forms* possible into which any System may be reduc'd, an Infinity of *Places* in which Animals may be situated, and an Infinity of *Relishes* or *Senses* in these Animals is suppos'd possible; that in the immense Spaces any one Animal should by Chance be plac'd in a System agreeable to its Taste, must be improbable as *infinite* to *one* at least: And much more unreasonable is it to expect from Chance, that a multitude of Animals agreeing in their Sense of *Beauty* should obtain *agreeable Places*.

In other words, since there is an unlimited set of possible attributes we could find beautiful and an unlimited number of possible worlds in which we might exist. The odds that not just you and I, but billions of people should by chance find that their tastes match the world in which they live are so small as to be utterly unbelievable. It cannot be an accident that our capacity to experience beauty is adjusted to the actual world in which we live.

This is really a rather sophisticated version of the argument from providence. But like the old Stoic argument, it relies in part on the implicit assumption that there are no possible causes of our happy state of affairs that are non-mental, non-random. While Hutcheson had no particular reason to doubt this claim, he also had no justification for it. Once again, the modern reader may be tempted to invoke an evolutionary explanation. However, in the case of beauty, such an explanation is not likely to be straightforward. It is not obvious why a sense of beauty is especially advantageous. One might argue that it is instead a by-product of an evolutionary advantageous trait.[26] But such a claim remains contentious.

---

[26] See, e.g., Dutton 2009.

7 Hume

7.1 Overview

Ask a modern philosopher about Hume and the design argument, and there's a good chance you'll hear this story: the design argument is an argument by analogy that compares the universe to a machine and concludes that, like a machine, the universe has a designer. Of course, that designer is supposed to be God. For a brief period of time following the scientific revolution, this argument flourished. But in the late eighteenth century, a philosopher named David Hume unmasked the design argument as a hopeless failure. Those that continued to defend the argument after Hume, most famously a fellow named William Paley, were merely propping up a corpse – the design argument had already died at the tip of Hume's pen.

As we will see in this chapter and the next, most of this tale is false. While Hume did in fact devastate the argument by analogy, it was largely an argument of his own creation. There are other types of design argument that, while seriously challenged by Hume, are not obviously defeated by him. In fact, Hume provides some of the most important rebuttals to his own critique, rebuttals that others like Paley will use to keep design arguments alive. But we're getting ahead of ourselves. Let's first take a moment to introduce Hume and the book that had such an impact on natural theology.

7.2 Who was Hume?

David Hume is by many lights the most important philosopher to write in the English language. His empiricist epistemology and the problems it posed for the justification of causal reasoning continue to exercise and vex philosophers to this day. Not only did Humean empiricism give rise to the logical positivism which dominated early twentieth-century philosophy in Britain and America, but the idealism of Kant – which marked the birth of a movement that would end in phenomenology in continental Europe – was a direct response to the problems posed by Hume. We could spend years discussing Hume's work and the various responses to it, but our interest in his work is relatively narrow. In the next section, I'll give enough of a gloss of Hume's epistemology to make his discussion of natural theology intelligible. In the meantime, let me situate the man in his proper time and place.

Hume was born in Edinburgh, Scotland in 1711. Though of high standing, his family was not wealthy. Hume's father died when he was an infant. So when Hume was old enough, his family encouraged him to study law. But the young Hume found he had an "insurmountable aversion to every thing but the pursuits of philosophy and general learning."[1] He ultimately abandoned his legal studies. With a view to avoiding poverty, Hume tried his hand – briefly – at business in 1734. This he also found "totally unsuitable" to his constitution. And so he directed his attention elsewhere. He retreated to France and, over the next three years, wrote his first significant philosophical work. The work was entitled *A Treatise of Human Nature*, and is today held in great regard. As Hume describes it, however, the book initially "fell dead-born from the press, without reaching such distinction, as even to excite a murmur among the zealots."[2] However, his next work, *Essays, Moral and Political*, was quite successful, and prompted him to rewrite the *Treatise*, resulting eventually in the *Enquiry Concerning Human Understanding* – the pithy statement of Hume's empiricism and of the problem of induction[3] that is still read widely today. Hume also wrote a very successful pair of texts on the history of England. The book which concerns us as students of design arguments – the *Dialogues Concerning Natural Religion* – was written before 1752 (the year Hume began composing his histories of England), but was not published until 1779, three years after Hume's death.[4]

7.3 Concept empiricism: a bit of epistemology

To make sense of the *Dialogues*, it will help to have at least a sketch of Hume's epistemology as it is presented in his *Enquiry*. To begin with, Hume

---

[1] Hume 1777, 4.   [2] Hume 1777, 7–8.

[3] For a concise discussion of Hume's Problem of Induction, see section 2 of Vickers 2009.

[4] Copleston 1993, Vol. V, 259–60.

calls the objects of thought 'perceptions'. There are two types of perceptions: impressions and ideas. Impressions are the lively ideas of sensation,
the mental objects that result from immediate sensation. For instance,
looking at this page gives me the impression of whiteness and the impression of little black typographical characters arranged in rows. Ideas by
contrast are less forcible and less lively; they are whatever elements of
thought we have that are immediate consequences of perception. These
include such things as memories, fantasies, and abstractions. All ideas do,
however, derive from impressions; each idea, says Hume, can be decom-
posed into a collection of simple ideas each of which is a copy of a simple impression. The image I have in my head when I attempt to recall
my children's faces though they are not at present in sight is an example
of an idea. As evidence for the claim that all ideas derive from impressions, Hume points out that when people lack a particular mode of sensation – such as vision in the blind or hearing in the deaf – they cannot have
certain types of impression. In these cases, we find that they also lack a
corresponding set of ideas. So, for instance, someone who is congenitally
blind will lack an idea of whiteness.

Ideas are connected to one another in the sense that they do not succeed one another at random in our thoughts. The succession of ideas
occurs according to a few basic principles which Hume calls resemblance, contiguity, and cause and effect. According to the resemblance
principle, an idea of, say, a particular automobile may be succeeded by
an idea of your own car. By contiguity, Hume means that an idea of the
beach can evoke an idea of the ocean – they are contiguous in space. The
most important principle by which ideas succeed one another, however,
is that of cause and effect. Here, the idea of a cause, such as striking a
match, leads us to call up an idea of the effect, such as the ignition of a
flame.

With this account of the origin and succession of ideas in hand, Hume
classifies all reasoning – the establishment of belief – into what he calls
'relations of ideas' and 'matters of fact'. Relations of ideas are roughly
those propositions which we would call a priori – they are discoverable
by the mere operation of thought, without dependence on what is anywhere existent in the universe. Reasoning about relations of ideas produces propositions that are certain (either intuitively or demonstratively)
and that are what Hume calls knowledge. Mathematical proofs are good
examples of reasoning about relations of ideas. As long as you have the

ideas of triangle, angle, side length, etc. then it follows with certainty that the interior angles of a triangle sum to 180°.

Reasoning about matters of fact, on the other hand, concerns the existence of objects or events beyond the present testimony of our senses or the records of our memory. The contrary of every matter of fact is still possible, and so this sort of reasoning results in conclusions that are only probable. Hume calls these beliefs. Most of our actual reasoning, both mundane and scientific, is reasoning about matters of fact. According to Hume, all reasoning concerning matters of fact is founded on the relation of cause and effect – we only infer what we do not currently see by using the cause–effect relationship. There is nothing about the idea of a cause that allows us to deduce its effect – if there were, it would be a relation of ideas! So all cause and effect relations can only be established through experience. For instance, it is only through experience that we learn to expect nourishment when we consume bread. This is not something we could have reasoned out in advance based solely on the idea of bread.

Arguments from design involve reasoning about matters of fact. The point of such an argument is to infer the likely existence of an unobserved God – whose non-existence poses no logical contradiction – on the basis of observed facts relating to design. For Hume, this means an appeal to experience and the relation of cause and effect. Thus, Hume's account of how we come to have knowledge of matters of fact is important for understanding the positions he takes in the *Dialogues*. Unsurprisingly, many of Hume's arguments are compelling only if we accept his view of epistemology and his account of our knowledge of cause and effect.

7.4 Setting the stage: Part I of the *Dialogues*

The *Dialogues Concerning Natural Religion* offers just what the title advertises: an examination of the consistency and soundness of the a posteriori arguments for the existence of a Christian God in dramatic dialogue form.[5] In these dialogues, Hume presents a variety of criticisms that are generally considered to have devastated the analogical argument. However, it does

---

[5] The term 'natural religion' generally denotes the effort to rationally justify or reason out the proper content of religious belief. This is to be contrasted with 'revealed religion', in which beliefs are fixed by divinely received or inspired revelations. The term 'natural religion' then really encompasses all arguments for the existence of God that

not seem that Hume is arguing in favor of full-blown atheism. Rather, he seems ultimately to want to deny a meaningful distinction between the position of theists who assert that the world was created by a mind bearing only a remote likeness to our own and the skeptics who insist that the order we see is the product of a principle that again bears some resemblance to human thought. I'll have a little more to say about what Hume's own position may have been and what we should take away as the conclusion of the *Dialogues*. But it should be said in advance that Hume does seem to think that his arguments conclusively block any a posteriori inference to a God having all of the attributes granted to him by revealed religion.

In writing the *Dialogues*, Hume was consciously imitating Cicero's work on the same subject, *De natura deorum*. The central character of Philo is "a modern-day Cotta, given the name of the skeptical head of the Academy who had successfully converted Cicero to his school."[6] Philo certainly plays the role of the skeptic, emphasizing the fallibility and limitations of human reason. His principal opponent in the *Dialogues* is Cleanthes, who presents arguments along Stoic lines in favor of religious belief founded on empirical facts. Fittingly, the character is named for the second head of the Stoic school (the successor to Zeno). Finally, there is Demea, a proponent of revealed religion and an advocate of the primacy of faith and reason above experience in ascertaining knowledge. He is named for a character in a play by the Roman comedian Terence, entitled *Adelphi* (The brothers). The play concerns the proper mode of education, and contrasts the laissez-faire approach of Micio with the authoritarian, traditionalist stance of his brother Demea.[7]

The *Dialogues* are framed as a letter from a young man named Pamphilus to his friend Hermippus. Pamphilus is a student under the charge of Cleanthes, and the conversation Pamphilus is relating to his friend begins as a discussion over how best to manage the religious education of children (presumably this concern with education prompted Hume's reference to the play of Terence). Demea compliments Cleanthes for his methods, explaining that he teaches natural religion last, preferring to 'season' his pupils' minds with proper religious practice and piety and to inculcate a sense of

do not depend either on authority or personal revelation, whether the arguments be a priori or a posteriori. Hume, however, considers only a posteriori arguments in his *Dialogues*.

[6] Cicero 1998, xliv.   [7] Terence 1886.

the absurdities to which philosophy can lead before exposing them to philo-sophical proofs of God's existence. In other words, it's dangerous to encour-age students to ground their religious beliefs in argument before they know what those beliefs should be. Philo seconds the notion, extolling the virtues of philosophical skepticism. He claims that to ground religion it is first neces-sary to debase reason. Cleanthes rankles at this claim and asserts that it is both possible and desirable to rationally justify religious belief. He argues that both the nature and existence of God are accessible to reason operat-ing on the products of experience. This initial disagreement launches the characters into an extended debate over the limits and conclusions of nat-ural religion in which Cleanthes defends natural religion, Demea defends revealed religion and dismisses Cleanthes' project, and Philo claims to be a skeptic out to show how reason generally leads to absurdity.

7.5 Hume's presentation of 'the' design argument

Hume gives us two distinct design arguments, though he is not careful to distinguish them. The first is analogical. The character Cleanthes asks us to look around the universe and note that it is essentially a machine – everywhere you look you see the adjustment of means to ends, the align-ment of structure with obvious functions. As he puts it:[8]

> The curious adapting of means to ends, throughout all nature, resembles exactly, though it much exceeds, the productions of human contrivance; of human design, thought, wisdom, and intelligence. Since therefore the effects resemble each other, we are led to infer, by all the rules of analogy, that the causes also resemble; and that the author of nature is somewhat similar to the mind of man; though possessed of much larger faculties, proportioned to the grandeur of the work, which he has executed.

Hume does not leave the argument in this form, however. The character Philo wants to clarify the argument before attacking it (a good philosoph-ical practice!). Here is a how he restates the argument, supposedly in a more precise form:[9]

(1) Anything which can be clearly conceived must be possible.
(2) Merely experiencing the state of the universe or of the order, arrange-ment, and adjustment of the final causes within it is insufficient to

---

[8] Hume 1990, 53.   [9] Hume 1990, 56–57.

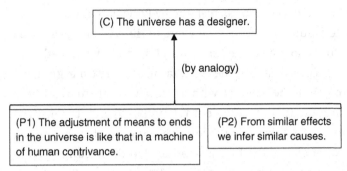

Figure 7.1   Cleanthes' analogical argument diagramed

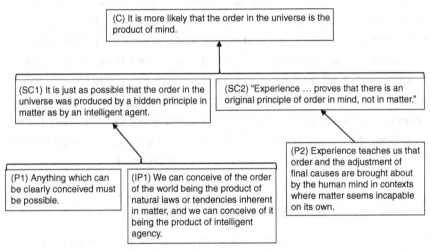

Figure 7.2   An argument from order diagramed

prove design – it is just as possible that matter is possessed of a hidden principle that brings all this about.

(3) However, experience teaches us that order and the adjustment of final causes is brought about by the human mind in contexts where matter seems incapable of doing so by itself.

(4) "Experience, therefore, proves that there is an original principle of order in mind, not in matter."[10]

(5) From similar effects, we infer similar causes.

(6) The adjustment of means and ends in the universe is like that in a machine of human contrivance (likewise for order, etc.).

(7) Thus we infer a designer of the universe.

---

[10] Hume 1990, 57.

The problem is that the argument as restated by Philo is really a composite of two separate arguments, though nowhere in Hume's *Dialogues* is this acknowledged. This is not an isolated slip. In fact, it might be argued that, as in the work of Cicero which Hume is imitating, these two arguments run together throughout the *Dialogues* without sufficient care to tease them apart. The first argument is captured entirely in (5)–(7) and it is just the analogical argument Cleanthes has already given, and which is diagramed in Figure 7.1. So what of the first four propositions in the argument Philo gives? They aren't necessary for making the analogical case; in fact, Philo's version of the argument by analogy is no better than Cleanthes'. Rather, these extra propositions make up the bulk of an argument from order that is diagramed in Figure 7.2. What distinguishes Hume's version of the argument from order from that of the Stoics is that in Hume's case the argument is overtly inductive. Since the argument from order involves an inference about matters of fact, the conclusion can only be established with a probability less than certainty. How much certainty is warranted depends on how much confidence experience gives us that order comes only from minds. We'll come back to this argument from order shortly. For now, let's look closer at Hume's analogical argument.

7.6 How analogy works

It is worth asking in general how convincing an argument by analogy can be. How can we decide just how much a conclusion is supported by noting that one thing is like another? Insofar as analogical reasoning is inductively strong, the strength of any conclusion to be drawn from an analogy depends on the degree of similarity of the cases at hand. To make this clearer, let us introduce some terminology.[11] When we say two things, events, or situations are analogous, we are saying that they share some properties in common. We'll call these the *positive analogies*. However, since the two things being compared are not the same thing, we know in advance that there are properties they do not share. We'll call these the *negative analogies*. For example, Bernoulli modeled the molecules of a gas as tiny billiard balls careening around at high speed. That is, he asserted an analogy between gas molecules and billiard balls. Clearly, there are

---

[11] I'm following the terminology used in Hesse 1964.

numerous negative analogies. We know that unlike billiard balls, gas molecules are invisible. They are not covered in paint. They do not weigh nearly as much. On the other hand, they do carry momentum, they collide elastically, and they are (often) roughly spherical. These are positive analogies.

Now, the point of drawing an analogy, at least so far as analogical argument goes, is to extract a fresh conclusion concerning a property that the two things being compared are not known to possess in common. So, for instance, I might want to know whether gas molecules are subject to gravitational attraction. Given the number of positive analogies between gas molecules and billiard balls, I might argue that there is good reason to suppose gas molecules to be subject to gravity just like billiard balls. That is, if I know two things are similar in certain respects, I can argue on that basis that they are similar in others. The strength of my argument, however, will depend on just how many positive analogies I can point to. So, for instance, a raven is like a writing desk in that they are both massive. If I was to use this analogy to argue that writing desks can probably fly, I would be on thin ice. The problem is that there are lots of negative analogies and few positive ones in the case of ravens and writing desks. So, for the proponents of natural theology to make a compelling case through analogy, they would need to show that there are lots of ways in which the universe is like a machine.

We'll see in the next section what Hume has to say about the plausibility of this analogical argument and arguments from analogy in general. But before we adopt a critical stance, it's worth showing analogies in their best light. Arguments by analogy abound. You and I employ them all the time whenever we draw conclusions about how likely we are to enjoy a film or food based on similarities with past movies or dishes we've liked or disliked, or whenever we infer the amount of time a task will take (like changing the oil in your car) on the basis of characteristics it shares with other tasks (like changing the coolant). Often we attempt to settle disputes by appealing to analogy with cases on which we agree. For instance, I might try to convince a child that his unwillingness to share a toy is unfair by pointing to the similarities between the current situation and one in which another child refused to share with him. Analogical arguments are also everywhere in science. To pick one sort of example, paleontologists would have much less to tell us about extinct species if they didn't make use of analogical reasoning. A common strategy when attempting to infer how a fossil animal looked or lived is to compare it with living species that are very similar. For example, if the teeth of your fossil animal have many of the features that are characteristic of mammals, then you would likely conclude that the animal was warm-blooded like a mammal too.[12] The point is that there is nothing obviously illicit about reasoning by analogy. Whatever defects the analogical design argument may have must involve the application of analogy, not the use of analogy per se.

7.7 A dilemma for the analogical argument

Speaking through the character of Philo, Hume objects to the analogical design argument by constructing a dilemma. That is, he sets out to demonstrate that the premises of the analogical argument for God force a natural theologian like Cleanthes to accept one of two distasteful conclusions. On the first horn of the dilemma is the weakness of the conclusion one can draw in this case based on the rules of analogical reasoning. Philo, parroting Hume's own epistemology, declares that all inferences concerning matters of fact are founded on experience, and that all 'experimental' reasonings are based on the supposition that similar causes produce similar effects and vice versa. However, he stresses that the inference to similar causes from similar effects *is only as strong as the similarity in effects*: "Every alteration of circumstances occasions a doubt concerning the event; and it requires new experiments to prove certainly, that the new circumstances are of no moment or importance."[13] To cast this in overtly analogical terminology, the strength of the conclusion drawn from an analogy depends on the relative proportion and overall number of positive analogies. So if we are to infer similar causes upon observing similar effects, we can only do so with confidence if there are lots of positive analogies between the effects and very few negative analogies. Herein lies the problem. If like effects prove like causes, then the more grand and complex the universe is, the more this 'effect' differs from human artifacts, and the weaker the conclusion of like causes becomes. The more impressive is the supposed handiwork of God, the less confidence we can place in the conclusion that it is in fact the handiwork of God!

---

[12] Benton 2010.   [13] Hume 1990, 57.

Suppose we ignore this problem and accept that there is in fact a strong analogy between human artifacts and the universe. Then we get stuck on the other horn of the dilemma. If in fact the similarity between the universe and human artifacts is sufficient to establish that they share similar causes, then a number of distasteful consequences follow. First, one cannot attribute infinite attributes to the deity. Since inferred causes must be proportioned to the observed effects, and observed effects are always finite, so too must be any inferred cause. Put another way, human intellect is eminently finite and so must be any cause remotely similar to it; if we want to infer an intelligent designer as an additional positive analogy, then it must be a finite designer. Second, one cannot ascribe perfection to the deity, even if it is granted that She is finite. We simply have no grounds for asserting that this is the best possible universe, and our experience would lead us to believe that there are lots of ways of improving it. Again, even if we grant that the universe *is* perfect, we cannot attribute the perfection to the workman. After all, in our experience such finely crafted things as ships can be the product of trial and error by very many very limited laborers.[14] This observation in turn prompts another concern: the argument from analogy cannot establish the unity of the deity. Since, in experience, complex designs are invariably the product of multiple cooperating intelligences, the analogical argument must give greater weight to the supposition that there are many gods. Finally, it is true of all things we know of that possess minds that they reproduce, are mortal, etc. All of these attributes then must be ascribed to the inferred cause. The analogical argument from design – if successful – forces one to be a perfect 'anthropomorphite', a fancy term for one who gives God humanlike attributes.

Philo presses the attack on the analogical argument further by restating the form of analogical argument: "[W]here several known circumstances are observed to be similar, the unknown will also be found to be similar" (Hume 1990, 80). Philo defends the Stoic account of the universe as a more plausible conclusion of the analogical argument. He notes that the organization of the universe is more like our bodies than our artifacts. Since we know of no minds without bodies, it would seem most reasonable to

---

[14] This observation is striking in its similarity to the mechanism of natural selection which Darwin will propose a century later.

conclude that the universe *is* a body (rather than an artifact) possessed of a mind which we identify with the mind of God. Suffice it to say, this is not where the typical eighteenth-century defender of natural theology wants the argument to end up. But it gets worse. Taking it for granted that he has already demonstrated that the universe resembles a living organism more than it does a machine, Philo notes that we must then attribute it the same sort of cause – organic generation. So if Cleanthes wants to embrace an argument by analogy, he is stuck either acknowledging the weakness of the argument for God, or accepting that in fact the universe is like an organic body produced in a process a lot like biological reproduction – Cleanthes must choose between a poor argument for the existence of the Christian God, or a stronger one for the non-existence of the Christian God. It seems the analogical argument is hopeless.

7.8 Hume's critique of the argument from order

Various scholars, particularly philosophers, insist that the design argument died with the publication of Hume's *Dialogues*.[15] There is something of a consensus that, following the objections posed by Hume through Philo, one could not rationally sustain support for arguments that purport to infer God's existence from order, purpose, or contrivance in the universe. For instance, Robert Hurlbutt says that Hume's "refutation of the design argument and of natural theology is logically devastating and conclusive."[16] This is only partly true. It does seem to be the case that Hume undermined the argument by analogy. What often goes unmentioned, however, is that Hume largely invented it as well. One would be hard pressed to find an explicit argument from analogy (rather than an argument from order or purpose) in the writings of any of the major proponents of natural theology. So before we can call Hume's critique a total loss for natural theology, we have to inquire after the other types of design argument.

Of the non-analogical varieties we've considered so far, Hume has the most to say about the argument from order, which is diagramed in Figure 7.2. Against this argument, Hume – again through Philo – offers a

---

[15] See, e.g., Sober 2005b, 132, and Ratzsch 2008.
[16] Hurlbutt 1965, xiii–xiv.

number of objections. First, he points out that intelligence, thought, and design as they are instantiated in human minds are together only a small fraction of the causes in operation in the universe. He questions whether it makes sense to attribute a cause of a part to the cause of the origin of the whole. Hume doesn't elaborate much about this worry, but there are some deep questions in the philosophy of science lurking here. We might put it this way: physical causes are always formulated against a background of events. The production of motion by force only makes sense if we are talking about the behavior of a portion of the universe with respect to the remainder. It is dubious whether one can make sense of the claim that the entirety of the universe or all motion within it is the effect of a 'force'. But even granting that this is plausible, says Hume, we should not choose such a minor and weak cause as intelligence. In our experience, the greatest effects are produced by non-intelligent causes.

Hume's second objection goes right to the heart of the matter with respect to arguments from order: how can we establish as a general, contingent truth that certain kinds of properties can only be produced by intelligent agents? How can we show that order comes from mind and mind alone? Hume claims that all arguments from experience require repeated observations of the associated cause and effect. But the universe, so far as we know, has occurred only once. Thus, we can establish no cause and effect relationships concerning its origin.

This objection is developed further after Demea, the traditionalist, takes offense at Philo's suggestion that the universe was produced by organic reproduction. When Demea demands to know what evidence there could be for such a claim, Philo explains that this is precisely his point; there is insufficient data available to distinguish amongst a very large number of plausible hypotheses for the origin of the universe and of the curious properties that drive our intuitions of design. He points out that there are many processes with which we have experience that produce the sorts of properties on which Cleanthes rests his argument. Order, apparently purposeful behavior, complexity, etc. are all produced in varying degrees by processes of reason, it's true, but also by instinct, organic 'generation' (animal reproduction), and 'vegetation' (plant reproduction). Since all of these are equally compatible with the empirical evidence, it is mere bias to select design. In fact, says Hume, it is not possible to establish on empirical grounds that order follows only from design. The reason is that there will always remain a class of phenomena, in particular the very phenomena to which Cleanthes appeals such as the structure of the universe as a whole, for which we cannot observe the origin or production. As Philo puts it:[17]

> You need only look around you ... to satisfy yourself with regard to this question. A tree bestows order and organization on that tree, which springs from it, without knowing the order ... And instances of this kind are even more frequent in the world than those of order, which arise from reason and contrivance. To say that all this order in animals and vegetables proceeds ultimately from design is begging the question; nor can that great point be ascertained otherwise than by proving a priori, both that order is, from its nature, inseparably attached to thought, and that it can never, of itself, or from original unknown principles, belong to matter.

To reinforce his claim that there are other conceivable sources of the order in the universe just as plausible as an intelligent deity, Philo develops a mechanical account of the origin of the ordered universe. While it obviously inherits a lot from the Epicureans, perhaps the most interesting aspect of this proposal is the sketch of a process that looks a lot like natural selection. To describe Philo's mechanical hypothesis, it will help to note that Hume and his contemporaries thought of the universe as an island of matter in a great void, possibly full of other such 'universes'.[18] Suppose the universe is such that matter is in perpetual agitation. If it were to begin in a chaotic state, this agitation would continue to randomly explore new configurations. If it should stumble upon a configuration in which we currently find the world, order would be maintained while matter remains in flux. As Philo puts it: "But is it not possible that it may settle at last, so as not to lose its motion and active force ... yet so as to preserve a uniformity of appearance, amidst the continual motion and fluctuation of its parts?"[19] That is, once it stumbled on to a dynamically stable configuration, the universe would be stuck there. Aside from natural selection, this resembles the sorts of arguments employed in statistical mechanics. The point is that, from a random beginning and purely material causes, the universe would come to perpetuate in a state that obviates the need for explanation of such things as the adjustment of animal parts one to another – in a dynamically stable state, everything would have to be adjusted for the

---

[17] Hume 1990, 89.   [18] Munitz 1981.   [19] Hume 1990, 94–95.

perpetuation of the whole, otherwise it wouldn't be stable! The apparent purpose in the interactions among parts of the universe is accounted for by the fact that the only dynamically stable configurations involve such interactions.

7.9 Rebuttals

Hume provides some serious counterarguments to Philo's attacks. These appear relatively early in the dialogue (Part III), and are supplied by the character Cleanthes. The main strategy of the counterarguments is *reductio ad absurdum* (or 'reductio' for short). This is an argumentative form in which one demonstrates that a set of premises leads to absurd and obviously false conclusions. From this, it follows that one or more of the premises is false. Cleanthes aims to demonstrate that the premises Philo used to argue against the design argument – namely Hume's own claims about establishing matters of fact – lead to absurd and patently false conclusions, and therefore Philo's argument must be unsound. To do this, Cleanthes offers two examples. First, imagine that an articulate voice thundering from the heavens is heard simultaneously around the world, speaking wisely to all peoples in their native tongue. Cleanthes says rather plausibly that only a fool would doubt the role of an intelligent agency. On the other hand, there is a strong disanalogy between the voice in the sky and human voices. According to Philo's principles, this should prevent us from concluding a similarity in the causes – we would be blocked from inferring a designer.

The second example involves a more elaborate sci-fi fantasy. Imagine, says Cleanthes, a world different from ours in two ways. First, there exists a universal human language that we can all read and understand. This is not so far-fetched. After all, he says, various non-human species share a common language, albeit a primitive one. Think of the squeaks and barks all prairie dogs use to communicate (my example, not Hume's). Second, in this world books reproduce themselves like living beings. In other words, your library is populated by books that are the offspring of other books. Suppose further that each book contains an articulate and perhaps even wise or masterful text in the universal language similar to the plays of Shakespeare or the epics of Homer. Upon looking inside the books and discovering this fact, it would be absurd, says Cleanthes, to deny that they had a cause bearing a strong analogy to mind and intelligence. Nonetheless, they show even less contrivance than living things. Thus, if Philo is correct we must reject an intelligent cause for the books. This is absurd, so Philo must be wrong.

This last rebuttal applies just as well to Philo's complaints about the argument from order. If we are precluded from witnessing the origin of text in the books – if we did not see how the self-reproducing books first came to contain such text – then according to Philo we cannot know that such order was produced by mind. Perhaps there is some natural principle which causes the text. But in this case, that objection sounds strained and a little ridiculous. Surely there could be no doubt that whatever cause resulted in the texts was intelligent in a manner similar to human minds.

It is a significant if puzzling fact that Hume never responds to these rebuttals. He has the narrator, Pamphilus, explain that Philo was "a little embarrassed and confounded," and "hesitated in delivering an answer."[20] No answer is ever given – Demea interrupts and changes the subject. So we are left wondering what Hume's position might be concerning the criticisms he put in the mouth of Philo.

7.10 The problem of evil for design arguments

The problem of evil is usually raised as a puzzle for orthodox Christianity: if God is infinitely powerful, infinitely good, and infinitely wise, then why is there evil in the world? Either God knows about it and won't do anything, in which case He is not infinitely good, knows about and can't do anything, in which case He isn't infinitely powerful, or doesn't know about it, in which case He isn't infinitely wise. So important is this puzzle that proposed solutions to it are called by a special name, 'theodicy'. But the problem takes on new urgency in natural theology, where the aim is to infer the properties of God from observable features of the world. If the world is full of evil, then it seems we have to attribute such a quality to God. In order to consistently apply the analogical argument, Cleanthes makes a final concession, one that seems unavoidable in applications of the analogical argument:[21]

[20] Hume 1990, 66.   [21] Hume 1990, 113.

If we preserve human analogy, we must for ever find it impossible to
reconcile any mixture of evil in the universe with infinite attributes;
much less can we ever prove the latter from the former. But supposing the
author of nature to be finitely perfect, though far exceeding mankind;
a satisfactory account may then be given of natural and moral evil, and
every untoward phenomenon explained and adjusted.

That is, the analogical argument forces us to embrace the finite nature
and capabilities of the deity in order to preserve his moral goodness.

But now the door is open to a stronger objection: the evidence, says
Philo, opposes a finite God who is both good and powerful. Philo's argument runs as follows. There are four circumstances which are responsible
for the greatest fraction of ills in the universe:

(1) Pains are employed to motivate animals to do what is in their best
    interest, when it seems that pleasures alone would have sufficed to play
    this role.[22]
(2) The world is conducted consistently in accord with general laws, when
    a powerful being could have done otherwise and intervened more fre-
    quently for purposes of removing evil.
(3) Organisms are only narrowly adapted to survive – a large perturbation
    in the environment results in suffering, pain, or death.
(4) While many parts of nature, such as wind, are useful, they very
    often stray outside the bounds of usefulness and become destruc-
    tive or chaotic. In this sense, the 'machine' of the world seems to be
    roughly made.

None of these circumstances appears to be a necessary characteristic of
existence; it is conceivable that the world might have been made differently. So what can we infer from them concerning the designer? Philo
argues that four options exhaust the possibilities: "There may *four* hypotheses be framed concerning the first causes of the universe: *that* they are
endowed with perfect goodness, *that* they have perfect malice, *that* they
are opposite and have both goodness and malice, *that* they have neither

---

[22] Interestingly, William Paley will attempt to turn this point on its head in his *Natural
Theology*. He argues that God must be beneficent because he gave us pleasure, a gift
above and beyond the requisite sensation of pain. Paley's move here is dubious – one
would require an argument that pain is somehow a physiological necessity for life
and that pleasure is thus superfluous. He offers no such argument.

goodness nor malice."[23] Since the evidence of experience is mixed, show-ing at best both malice and beneficence, we cannot support the first two options with an empirical argument. The uniformity and steadiness of the general laws which guide the universe suggest that there is no conflict of motive, so we can probably dismiss the third. This leaves the fourth option as the most probable on empirical grounds. The inevitable conclusion of a successful analogical argument, says Philo, is that there exists an utterly apathetic, finitely endowed designer.

7.11 Other criticisms and Hume's surprising resolution

Hume's *Dialogues Concerning Natural Religion* is a relatively concise work, but dense with arguments. Many of these I've ignored because they do not bear directly on design arguments, or because they are relatively minor. For instance, Hume provides a reductio against the so-called cosmological argument (Aquinas' 'First Way'). Roughly he uses the form of the cosmo-logical argument to demonstrate that there must exist a God responsible for the evil in the world. Suffice it to say this upsets Demea, who departs the conversation. Unfortunately, we haven't the space to consider these peripheral arguments with the attention they deserve. But I will digress just a little in order to share the conclusion of the *Dialogues*, something generally ignored in discussions of Hume's role in the history of design arguments.

In the concluding chapter of Hume's *Dialogues*, Philo brushes aside all objections, affirms his own belief in the Christian God, and describes a position that is plausibly closest to Hume's own. Of course, you have prob-ably realized by this point that it is less than obvious which character, if any, is intended to represent Hume's thoughts on the subject. While Philo is the counterpart of Cotta in Cicero's dialogue and thus a good candidate for Hume's avatar, Cleanthes is given some strong, unanswered rebuttals. Furthermore, Hume ends the *Dialogues* with Pamphilus declaring "that, upon a serious review of the whole, I cannot but think, that Philo's prin-ciples are more probable than Demea's; but those of Cleanthes approach still nearer to the truth."[24] We the readers are thus left to evaluate the

---

[23] Hume 1990, 122; emphasis in original.
[24] Hume 1990, 139.

arguments on their merits, uncertain of which character is Hume's mouthpiece. Nonetheless, given Hume's empiricist epistemology, the final maneuver made by Philo is very consistent with Hume's approach to a variety of philosophical problems. The empiricist trick of which I speak is to declare an intractable question meaningless. Recall that Hume is a concept empiricist. All ideas are composites of simple ideas copied from our sense impressions. If you cannot dissect the simple ideas that make up the idea to which a word refers, then that word refers to no idea at all – it is meaningless. In this case, Philo argues that the difference between the theist and atheist positions is in fact meaningless: there is no content to the question of which is right because, insofar as either position is meaningful, they mean the same thing. As Philo puts it, the theist allows that the original intelligence is very different from human reason. The atheist allows that the original principle of order bears some remote analogy to it. "Will you quarrel, Gentlemen, about the degrees, and enter into a controversy, which admits not of any precise meaning, nor consequently of any determination."[25]

Before assenting to this resolution there is much work to be done. Is it really the case that there is no difference in empirical content between the two positions? I won't attempt to settle this question here. But I will suggest that, before dismissing Hume's final account, you consider the nature of 'physical law' as we currently use the expression. Whatever laws are and in whatever way they may be said to be responsible for the origin and arrangement of matter, there seems to be something to the claim that they are remotely analogous to the human mind, if only for the simple reason that they may be comprehended and described by such minds.

[25] Hume 1990, 130.

8   Paley

8.1 Who was William Paley?

In the modern era, there is no bigger name in natural theology than that of William Paley. He ended Cicero's nearly 2,000-year reign as the principal authority on the subject. Prior to 1800 CE, Cicero was widely cited and widely emulated. Even Hume composed his critique as a dramatic dialogue that closely imitated Cicero's *De natura deorum*. By the 1830s, however, Cicero had moved to the footnotes and the work of William Paley had become the most ubiquitously cited, referenced, and quoted. There is little in Paley's biography to foreshadow this intellectual upset. By all accounts his life was uneventful. In 1743, he was born into relative comfort as the eldest son of the Reverend William Paley. The younger William was regarded as a bright but clumsy young man, fond of fishing. When he exhausted the resources of the local school, he was sent off to Cambridge University at the tender age of 16. There, like his intellectual forebears John Ray and William Derham, he was ultimately ordained in 1766.

When he was later elected fellow of his old Cambridge college, Paley returned to teach. It was as an instructor in moral philosophy that he began to stand out. His lectures were innovative and well received, and he drew upon them to write his first book, *The Principles of Moral and Political Philosophy* in 1785. The book attempts to ground a utilitarian ethics upon the will of God and was a great success. It was followed by another, *Horae Paulinae*, in which Paley tries to authenticate the narratives of the New Testament, in particular the story told by the letters attributed to Paul. It is somewhat ironic that his strategy in that book is to argue that similarities among the various biblical texts are *not* the product of design.[1] This book

---

[1] In the *Horae Paulinae*, Paley argues for a common authorship of the letters traditionally attributed to Paul and for the authenticity of these letters and the history known

was succeeded by *Evidences of Christianity* – a broader attempt to marshal historical evidence in favor of the faith. It was so highly regarded that by the time a young Charles Darwin was a student at Cambridge, it had been made part of the standard curriculum.

Paley's final and most enduring success came at the end of his life. Out of a desire to complete his series of arguments for the truth of the Christian faith, he took up the challenge of natural theology. In 1802, just a few years before his death, Paley published *Natural Theology, or Evidence of the Existence and Attributes of the Deity, Collected from the Appearances of Nature*. As the lengthy title suggests, the book is an attempt to establish the existence of God using the empirical facts of eighteenth-century science. It was this book that displaced Cicero as the primary reference. *Natural Theology* went through numerous editions as the nineteenth century progressed, quite a few of which were edited or illustrated for niche markets. It was tremendously popular with expert and lay audiences alike. Why this text and not one of the many others written around the same time rose to such prominence and overshadowed the works of prior millennia is an interesting historical and sociological question, but one that is beside the point when considering the viability of design arguments. What we're interested in is the argument.

8.2 Paley's design argument

Paley begins his exposition with a puzzle, cast in terms of the most famous image in all of natural theology:[2]

> In crossing a heath, suppose I pitched my foot against a *stone*, and were asked how the stone came to be there, I might possibly answer, that, for any thing I knew to the contrary, it had lain there for ever: nor would it perhaps be very easy to shew the absurdity of this answer. But suppose I had found a *watch* upon the ground, and it should be enquired how the watch happened to be in that place, I should hardly think of the answer which I had before given, that, for any thing I knew, the watch might

as the Acts of the Apostles. A forger, says Paley, would harmonize his fake letters with the history but do so in such a way that the obvious or important points align. His entire strategy for proving authenticity is to seek out coincidences among the texts of the sort that indicate a *lack* of design.

[2] Paley 1802, 1–2; emphasis in original.

have always been there. Yet why should not this answer serve for the watch, as well as for the stone?[3]

The question is not rhetorical; the two examples provide an important contrast. No one would deny that we can infer design in the case of the watch, so by carefully considering why we are justified in this inference, Paley hopes to discover an argument that can then be applied to controversial cases like living things or the universe as a whole. So what does Paley tell us about the argument? How do we infer design in the case of a watch? We make the inference, says Paley, from the fact that the parts of the watch (unlike those of the stone) are arranged for a purpose. More explicitly, he says that the parts:[4]

> are so formed and adjusted as to produce motion, and that motion so regulated as to point out the hour of the day; that, if the several parts had been differently shaped from what they are, or in any other order, than that in which they are placed, either no motion at all would have been carried on in the machine, or none which would have answered the use, that is now served by it.

The unstated premise seems to be that such arrangement for a use or purpose can only be the result of intelligent agency, and thus we can infer a designer.

Granting for the moment that we've identified a property – arrangement for a purpose – that supports an inference to intelligent agency, we can apply the argument with fresh premises to infer the existence of God. To supply those premises, we look not to human artifacts but to living things or, more specifically, to certain of their parts. I'll focus here on Paley's first and probably best-known example, the vertebrate eye. The many parts of the eye are exquisitely arranged for the purpose of vision. To be more careful, since, as Paley admits, we do not know precisely how the sensation of vision is connected with the associated physical phenomena, we should say that the components of the eye are arranged for the purpose of reliably producing a sharp and faithful image of the environment on the membrane of the retina. At first go, the eye accomplishes this effect in much the

---

[3] If this sounds familiar, that's because Paley 'borrowed' (without attribution) both this example and the bulk of his argument from Nieuwentyt. For a discussion of Paley's apparent plagiarism, see Anon. 1833; Anon. 1849; Blakey 1859, 222–36.

[4] Paley 1802, 2.

same way that a telescope produces a sharp and faithful image of a distant object on the eye of the observer. In a telescope, a series of refractive lenses are placed within a tube in front of the observer in order to focus distant light into a magnified image at the location of the observer's eye. In the eye itself, a series of transparent, refractive lenses are placed in front of the retina to focus light coming from objects into an image at the surface of the retina. At the front of the eye is the cornea, the transparent membrane that bulges slightly outward in the center of the eye. The round cornea refracts incoming light, coarsely focusing it and thus acting as an optical lens. Behind the cornea is an aperture of muscle – the iris – which is the colored part of the eye. Light passing from the cornea through the opening in the iris falls upon a round, crystalline body properly referred to as the 'lens' of the eye. Unsurprisingly, the lens of the eye acts as an optical lens, and is responsible for finely focusing the light that is coarsely refracted by the cornea into a sharp and faithful image on the retina.

There is much more to be said concerning the adaptation of the parts of the eye for the purpose of providing a sharp image at the retina. Paley provides a lot of detail about the different structures found in various kinds of animals. Many of these details we now know to be false, but the modern facts substitute just as well. I'll stick with the latter and provide a couple of examples of the sort Paley has in mind. First, lenses – whether made of glass or protein – tend to suffer from 'spherical aberration'. That is, light striking the periphery of the lens is focused at a different point from light striking near the center and this makes the overall image blurry. In some vertebrates such as fish, this is corrected by layering the optical properties of the lens such that light is more strongly refracted near the center than at the edges. Furthermore, the delicate tissues of the cornea – which must be delicate in order to possess the required optical properties – are protected by the eyelid and moistened by a system of lacrimal glands (which produce tears) and ducts (which dispose of the excess fluid). Unlike telescopes but very much like an auto-focusing digital camera, eyes function to produce a sharp image of objects that are sometimes close to the eye and sometimes far away. This means the eye requires a system for adjusting its focal length. In fish and amphibians this is accomplished by a muscular system that pulls the lens forward or backward in the eye. In humans and other mammals, it is accomplished by special muscles that tug on the periphery of the lens and flatten its shape. I could go on detailing the components of the typical vertebrate eye and the delicate manner in which they are adapted for reliably producing a sharp retinal image. Paley does in fact go on at much greater length in this first application of his argument. But for our purposes, it suffices to note that the eye, like the watch, has a lot of parts that are clearly arranged for a particular use. Thus, according to Paley, the eye must have a designer, and the only suitable designer is God.

The bulk of Paley's book consists of an enumeration of many such examples, from the structure of the human ear to my favorite recurring example, the drill-like ovipositors of wood-boring insects. Paley emphasizes that each of these examples alone supports a compelling inference to design and the existence of a grand designer on the basis of the argument he identified in the uncontroversial case of a watch. The argument is cumulative just in the sense that we may be wrong about the details of any one example. Having lots of examples ensures that the relevant premises really are true of one or more living things – we can be reasonably certain that purpose is present in the structure of at least some organisms. Thus, the more examples we accumulate the more confident we are that at least one supports a cogent or sound argument (depending on how we interpret the structure of the argument).

Paley thinks that our observations of the natural world entitle us to conclude much more than the mere existence of one or more designers – he argues from the scientific facts to the conclusion that there exists a *single* designer of all living things and, more tenuously, of all the world. You may recall that demonstrating the unity of the designer proved problematic for Aquinas and was one of the principal weaknesses of the argument by analogy that Hume attacked. So how does Paley argue for the unity of the designer? He does so by appealing to the unity of the plan. He points to the heavens and notes that a single principle of gravity coordinates all of the various motions. The light from the stars behaves in the same way as the light from a candle and likewise the heat thrown off by the sun is no different in kind or action from the heat of a coal fire. Looking to specifics, he notes that the planets all experience seasons and all possess atmospheres of one sort or another.

Similarly, if we turn to the biological world from which Paley has so far drawn the strongest examples, we again see a broad uniformity in plan. We have to be a bit loose with the notion of uniformity, but the idea is to note the many similarities in the body plans and anatomy of terrestrial vertebrates – a deer or a turtle has more or less the same bones and organ systems as I do, though they vary widely in detailed shape and function. Turning to the fishes, the similarities lessen but remain numerous. Fish, after all, have a stomach, a liver, a spine, etc. Things get a bit murkier when we look to groups of organisms like the insects. For them, Paley suggests, we should recognize a "law of contrariety"[5] – insects are the structural opposites of vertebrates. While we have our skeleton on the inside, theirs is on the outside. Our muscles pull against internal bones, their muscles pull against an external shell. In this way, says Paley, the anatomy of insects represents a "remembrance"[6] of the anatomy of vertebrates.

All of this uniformity in physical, astronomical, and biological phenomena suggests a uniformity of plan on the part of the designer. While the notion of similarity here may be implausibly loose in some cases, the intuition is clear. The unity of astronomy, of physics, and of biology suggests a single designer with a coherent plan – there is none of the conflict one would expect if multiple designers were at work. Paley also goes on to argue for other properties of God such as benevolence, but we won't take up those derivative arguments here.

By this point, I've only sketched the argument as Paley presents it – I have yet to do any work filling in premises or clarifying and strengthening the inferences. Nonetheless, there is one immediate objection we should consider. This is the objection from reproduction, and it runs something like this: for Paley's argument to work, it must be the case that the 'purpose' inherent in such biological structures as the eye most plausibly originate from an intelligent source. However, says the skeptic, I can tell you exactly where the eye comes from and it isn't an intelligent designer. The source of, say, the eye of a fish is just another fish. That is, at some time or other a male and female fish combined their gametes (sperm and egg) which resulted in an embryo. This embryo developed on its own, either within the female fish or externally in an egg until it emerged in the form we now find it, complete with eyes. The source of the structure of interest – the eye – is just another living thing.

Paley anticipates this objection and dedicates two entire chapters early in the book to addressing this concern. While presenting the argument for design in the case of a watch, Paley asks us to consider a bit of science fiction. Suppose, he says, that the watch we found out on the heath were in the

5 Paley 2006, 235.   6 Paley 2006, 236.

course of our observations to produce a second identical watch. It is not inconceivable that a watch might contain a tiny set of machinery – lathes and dies and such – so that as it keeps time it mechanically assembles another watch. Today's technology only makes such a possibility seem more plausible. If the watch could do such a thing, then we would justly infer that it too was likely produced by a similar watch in the course of its functioning – the watch on the heath was probably the mechanical offspring of a parent watch. In this case, Paley asks, would we deny that the watch we had found was the product of design? The skeptic could say, "Look, we know where the apparent purpose of its many parts came from. It came from the watch that made it!" Paley argues that this objection misses the point. If you found a watch on a heath and, while observing this watch, found it to replicate itself, you would only be more convinced that the watch was designed and further impressed by the skill of the designer. Why? Because pointing to the watch's parent as the source of its apparent purpose is to point to the cause of the wrong thing. Yes, in one sense we can explain why this particular collection of parts has the structure it does by appealing to another collection of parts that assembled it. However, Paley argues that this fails to answer the relevant question. What needs to be explained is not how this particular lump of matter came to be formed into a watch through time – we're not interested in the immediate history of the parts. Rather, what needs to be explained is why any watch at all – why something exhibiting this sort of arrangement of parts to satisfy a use – exists at all. If as a matter of history the watch was formed by another watch, that doesn't explain why a sequence of self-reproducing watches exist. It fails to account for the origin of the 'purpose' that is manifested in every generation of reproducing watches. Note that this is precisely the answer Hume's character Cleanthes gives to the same objection when it is cast in terms of self-reproducing books (see Chapter 7).

8.3 The argument as analogy

In the last section, I tried to sketch the argument in terms faithful to those used by Paley. As a consequence, I was circumspect as to how we should best understand the relevant premises and inferences. What exactly is Paley trying to say? How can we charitably reconstruct his argument for the existence of a designer? Despite his apparently clear prose, the form of Paley's argument is ambiguous. There are at least four ways in which

philosophers have reconstructed it, and we'll have to do some work to identify the strongest possibilities. The first approach treats Paley as offering a
straightforward argument by analogy of the sort considered in Chapter 7.
In this view, Paley's talk of watches and telescopes is crucial because these
serve as the models of an analogy that runs as follows:

Organisms are like watches (or machines in general) with respect to
properties $P_1, P_2, P_3 \ldots P_n$. [The P_i represent the 'positive analogies'.]

Watches have designers.

Organisms have designers.

But this interpretation is belied by Paley's remarks in his sixth chapter,
entitled "The Argument Cumulative." Here, Paley asserts that:[7]

If we had never in our lives seen any but one single kind of hydraulic
machine; yet if of that one kind we understood the mechanism and use,
we should be as perfectly assured that it proceeded from the hand, and
thought, and skill of a workman, as if we visited a museum of the arts,
and saw collected there twenty different kinds of machines for drawing
water, or a thousand different kinds for other purposes.

In other words, he is claiming that were we to encounter a hydraulic
machine that we had never seen before *and had no other experience of machines
on which to draw*, then we could still infer that the machine was designed
by noting some of its peculiar properties. But if we have experience of
only the one machine, we cannot draw an analogy with other machines
in order to infer design. So if Paley's claims about his own argument are
accurate, the argument cannot be an analogy. Furthermore, if Paley is
making an analogical argument he wastes an awful lot of time early in the
book arguing from the properties of a watch to the existence of a watch
designer. For the analogical argument, all we need to do is note that in our
experience watches have the property of being designed. Finally, there is
good reason to think that Paley was aware of Hume's critique of the analogical argument and was taking pains to avoid these objections.[8] For all
these reasons, we can safely reject the analogical argument as an accurate

---

[7] Paley 1802, 82–83.

[8] Various portions of *Natural Theology* have been interpreted by scholars as indirect
responses to the arguments Hume gave in his *Dialogues Concerning Natural Religion*.

representation of what Paley had in mind. We can also set it aside in our search for strong design arguments since, as we have already learned, the argument by analogy is quite weak. Fortunately, alternative readings of Paley offer some more promising approaches.

8.4 Inference to the best explanation

Another way of reading Paley involves a type of argument we have yet to consider. Arguments of this sort – which we'll call *abductive* arguments – have the following general form:

Premise 1: Surprising fact *C* is observed.

Premise 2: If proposition *A* were true, *C* would be likely.

Conclusion: *A* is true.

Abductive arguments are clearly not deductive since it is certainly possible for the conclusion to be false even if all the premises are true. Whether this sort of inference should count as an argument at all – whether premises of this sort really lend any support to the conclusion – is a matter of debate. Some scholars view abductive inferences as incapable of establishing *any* degree of confidence in the conclusion. In fact, the American philosopher Charles Sanders Peirce, who introduced the notion of 'abduction', saw it as performing a role distinct from either deductive or inductive inference. For Peirce, the two premises of an abductive argument only serve to establish that *A* is a viable explanatory hypothesis for the surprising fact *C*. If a hypothesis can be abductively inferred from a set of premises, then it can be said to explain those premises. One is not entitled, however, to any particular degree of belief in a hypothesis just because it follows from an abductive inference.[9] As Peirce saw it, abductive inferences are merely a means of identifying and ranking those hypotheses we should dedicate our limited resources to testing, not for establishing the probable truth of those hypotheses.

Let me give you an example. Suppose that, walking through a field in Kansas, I discover a cow in a tree. This is a surprising fact. Now, I might

Whether or not this is true, Paley explicitly cites Hume at the end of his book (Paley 1802, 265) and so at least had to be aware of the *Dialogues*. For further discussion of the influence of Hume on Paley, see, e.g., Gillespie 1990, 219.

[9] Peirce 1955.

hypothesize that a tornado passed through here not long ago. This would certainly explain the surprising fact, and so follows from a proper abductive inference. For Peirce, the inference licenses only the claim that the tornado proposition is a viable hypothesis. I might also hypothesize that the tree-borne cow was the victim of alien abduction. If aliens did abduct the cow then eject it from their spaceship, it would explain the surprising fact and is therefore also supported by an abductive inference from the evidence. If we add to our surprising fact that there is debris all around, corn flattened in the fields, and an RV wrapped around the base of the tree, then the tornado hypothesis seems to be more explanatory than the alien hypothesis. But, says Peirce, we cannot say that it is more likely to be true.

Many philosophers since Peirce have claimed that certain sorts of abductive inference are in fact viable ways of establishing the probable truth of hypotheses and are therefore really inductive inferences. For instance, if all competing explanations for a surprising fact can be eliminated or shown to be less probable so that a single 'best' explanation remains, then some think we are entitled to infer the probable truth of the remaining hypothesis. This sort of argument is often called an *inference to the best explanation*.[10] In this view, if I could show that aliens have never visited Earth in spaceships and that I have exhausted all other plausible alternatives, then I am entitled to a high degree of confidence in the tornado hypothesis. Inference to the best explanation is the only sort of abductive inference we'll see used in design arguments, and we'll assume that it is an acceptable sort of inductive inference.

8.5 Paley's argument as an 'inference to the best explanation'

As I mentioned in the last section, some philosophers view Paley's argument as an inference to the best explanation.[11] In the case of a watch discovered in a deserted meadow, some surprising facts include the precise adaptation of its many parts to the apparent purpose of reliably keeping time. To be a little more specific, the watch is found to contain a number of parts that are arranged in just the right way so as to produce a

---

[10] See Harman 1965.
[11] See, e.g., Sober 2000 and Shupbach 2005.

regular motion of the hands of the watch. If any of the parts were of a slightly different shape, in a slightly different position, or – in the case of the spring – of a slightly different composition, the watch would cease to exhibit this extraordinary behavior. Next we take stock of the available explanations for these surprising facts. For Paley, there are only two: (1) the watch was designed and constructed by an intelligent agent who wished to keep time, or (2) the parts of the watch were randomly tossed together by unthinking physical processes, rather like the 'dust bunnies' that form under one's bed.

In an inference to the best explanation, we take each hypothesis in turn and ask, "How likely would the surprising facts be if the hypothesis is true?" Let's assume, then, that there is some intelligent agent – perhaps a Swiss jeweler – who is interested in keeping time and is capable of constructing intricate machinery. Then it is quite likely that something with the surprising properties of the watch would be produced. On the other hand, if we imagine the elements – wind, flood, etc. – as having swept together the various parts, it seems extraordinarily unlikely they should be so well coordinated. Thus, design is the best explanation and we can infer that the watch most likely has a designer. Overall, the argument for the watch looks like this:

> Two possible explanations for the properties found in a watch (organism) are chance and design.
>
> The hypothesis of design makes the observed properties much more likely than they would be if produced by chance processes.
> _____
> The watch was designed.

We can represent the argument as a diagram like that shown in Figure 8.1. I'll stick to the convention of labeling the arrow indicating an inference to the best explanation "IBE."

In this view of Paley's argument, the watch example is supposed to illustrate the inference he wants to make with respect to living things. Like the watch, any given organism is rife with surprising features that cry out for explanation, especially those involving the remarkable adaptation of its structure for satisfying its specific needs. Just think of the various structural features of the eye discussed above that produce a sharp image on the retina and so provide the bearer of the eye with a means of

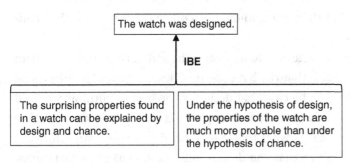

Figure 8.1   Paley's design argument interpreted as an inference to the best explanation

navigating and assessing its environment. Paley considers two hypotheses to account for these facts. First, we might suppose the parts of any given organism to have been assembled by chance from the random collision of atoms, much as the Epicureans envisioned. This, however, makes the outcome – a highly structured organism adapted for many functions – extremely unlikely. The facts grow increasingly improbable on the chance hypothesis the more organisms and the more adaptations of each organism we consider. Alternatively, there is the possibility that living things were designed by an intelligent agent with sufficient resources to do the job. In that case, it would be no surprise at all that organisms are adapted to their environments. Of the two, design is the best explanation.

Let's grant the cogency of this argument. In particular, let us suppose that design is the best explanation among those considered. How much confidence should we put in the conclusion of design? The answer depends in part on two things. Both suggest general difficulties with inferences to the best explanation. The first has to do with the fact that the 'best' explanation is always relative to a set of alternatives. If we add or subtract explanatory hypotheses, we can alter which rises to the top as the 'best explanation'. For instance, if we add evolution by natural selection as an explanation for the adaptation of organisms – a hypothesis we'll take up in the next chapter – then it is no longer clear that design is the best explanation. Certainly if we contrast natural selection and the chance hypothesis alone the argument would come out in favor of natural selection. Of course, Paley had no conception of natural selection, but that is precisely the point. A hypothesis we have yet to think of might trump all the others on the table. My confidence in the conclusion of design therefore extends no further than my confidence that I've thought of all the plausible hypotheses.

The second point relevant to assessing the strength of this argument concerns the premise that the hypothesis of design makes the surprising facts likely. To make the case that design is the best hypothesis we need to show that it does a better job of explaining the facts in question. We haven't said much about what that might mean, but one straightforward approach is to require that the design hypothesis make the facts more probable than chance alone would. But to know how likely it is that an intelligent agent would produce a particular outcome, we need to know something about both the agent's desires or goals and its capabilities. In the case of a watch, we know that the ability to mark out regular intervals of time is something humans often desire and we know that humans are capable of synthesizing machines of comparable complexity from similar materials. In the case of living things, Paley gives us no account of what sort of motives or capabilities the hypothesized designer might have. We might simply say, "Well the designer in question is one who would want to produce organisms like this and he would have the ability to do so." Such a hypothesis would make the facts very likely indeed. But if we allow into consideration hypotheses like this that simply work backward from the data at hand, then we trivialize inference to the best explanation. One could always concoct an explanation in terms of a designer or chance or anything else one likes by tacking on enough assumptions to make the facts at hand likely. For instance, I might take the chance hypothesis to be that organisms are the result of the blind collision of atoms *and* that the universe happened to occupy just the right initial state to make these collisions inevitable. Such a claim makes the observed facts likely even though it isn't very plausible. One might think we can just insist that all hypotheses considered in an inference to the best explanation must be plausible on grounds independent of the facts at hand. Cooking up criteria by which to assess independent plausibility is not straightforward and we cannot pursue the project here.[12] The important point is that if we want to be confident in our inference, we'll need to specify the designer's attributes, and we'll need some way to rule out hypotheses that 'cheat'. Note that neither of the above objections is fatal to the argument. However, both provide the seeds for powerful objections that we'll meet in coming chapters.

[12] For an attempt at providing such criteria, see Collins 2009.

8.6 The argument from order

The strongest reconstruction of Paley's argument – and that which I would argue is closest to his intention – is a deductive inference based on premises that are themselves inductively justified by experience. Viewed this way, Paley offers a version of the argument from purpose that runs like this:

Features of organisms such as eyes and ears exhibit purpose.

Purpose implies a designer.

Therefore, there exists a designer of living things.

This kind of argument was introduced in Chapter 3 when we considered the various Stoic positions, but we didn't pause to consider it in any detail. The argument from purpose as stated faces some serious difficulties. The problem has to do with how we understand the term 'purpose'. When we say that the properties of an object exhibit purpose, we might mean that the object has the properties it does because some intelligent agent intends it to accomplish some end. That is, a thing has purpose just if some intelligent agent intends that it accomplish something. So, for instance, my pen exhibits purpose because the engineers who created it intended it to satisfy the end of making marks on paper, and its properties were chosen to further that end. This is typically what we mean when we talk of purpose. When understood this way the premise that purpose implies a designer is true by definition – part of what we mean when we say a thing has a purpose is that it has a designer. But this means that we beg the question of design when we assume that organisms exhibit purpose in their structure. That would be like assuming that organisms were obviously crafted by a designer and then inferring the existence of a designer. This is a viciously circular argument, one which offers no support for its conclusion.

Writing in the late nineteenth century, the geologist Lewis Hicks diagnosed this mistake lurking in the argument from purpose (which he and others call the "teleological argument").[13] Hicks suggested that the reason the argument from purpose is so intuitively appealing to many people is that they unwittingly conflate the purpose of a thing with its function. By function I just mean what the thing does. By saying that an eye forms

---

[13] Hicks 1883.

an image at the retina (that an eye functions to provide sight) is not to say that the eye was intended by some agent to have that function. It is only when we are sloppy and assume a purpose when we see a function that we run into trouble.

Whether or not Paley makes the mistake Hicks points to, we can help him out by reading his use of 'purpose' in terms of function. How would the argument go in this case? The idea is to look at all of the things a system actually does. For each thing it does – each *function* – ask how many parts contribute and, of those that do, how many of their properties are relevant for producing that effect. Of the properties under consideration, what range of values would have allowed them to accomplish the same function? If many parts contribute to a function and small variations of the properties of most of those parts would preclude the function, then we can say that the system under consideration is *goal-directed* with respect to that function. That does not mean it was designed for the function. Maybe it was, maybe it wasn't. It's just to say that there is a special relationship between the complex structure of a thing and its behavior.

Consider Paley's example of the watch. With respect to the function of marking out regular intervals of time, many or most pieces of the watch are essential and each of them must have properties (like shape or stiffness) within a very narrow range of possibilities. According to the above account, that would make the watch goal-directed with respect to the function of marking time. For a biological example, consider the human heart. It has many functions. But with respect to the function of pumping blood – something the heart clearly does whether or not it was designed – many of the details of its structure and physiology are very relevant. For instance, if the chambers lacked one-way valves between them there could be no bulk motion of the blood. If there was no mechanism for coordinating contractions of the various chambers, then no blood would move and to get it to move as efficiently as it does requires very precise coordination. So with respect to pumping blood, details matter and there are many relevant details. In that case, we can say that the heart exhibits goal-directed behavior with respect to the function of pumping blood. Of course, not all of a heart's functions exhibit goal-directedness. Among the many behaviors it exhibits is the function of making noise. Relatively few of the heart's components are required for this function. For instance, the heart need not have a coordinated system of valves in order to make noise – it would make noise if the valves opened and closed at random. Similarly, the distribution of muscle in the heart would be irrelevant. The salient point is that at least some of the heart's functions are goal-directed in the neutral sense spelled out above.

Of course, this notion of goal-directedness is only useful if it picks out a special kind of system – if everything turns out to have goal-directed behavior then the concept is useless for constructing a successful design argument. To see that it really does exclude what we want to exclude, just think of the humble stone lying in the field. It doesn't seem to have any functions that satisfy the conditions for goal-directedness. Stones don't do much by way of expending energy but they do have functions. They exert a force on the ground beneath them. They conduct heat. Some of them are magnetic, etc. But for each of these functions, none of the stone's parts are necessary, and the parts could assume any of a wide range of properties – weight, density, conductivity, etc. – and still satisfy the same function. I have left many important details unattended, but this rough account of goal-directedness seems to do the job of picking out the kinds of systems with which Paley was concerned.

It will help at this point to step back and restate the argument with the suggested modifications. It now goes like this:

> Many systems in each organism are goal-directed.
>
> Goal-directed systems can only be produced by design.
> _____
> Therefore, there exists a designer of organisms.[14]

While this revised argument no longer begs the question, it no longer resembles the old argument from purpose either. Instead, what we have done by adjusting and precising Paley's language is to produce a version of the Stoic argument from order (see Chapter 3), where the special property of 'order' has been identified with goal-directedness. But recasting the argument this way introduces a new problem. The second premise – that goal-directedness implies a designer – is no longer true by definition. It is not part of the definition of goal-directedness that there exists a designer. So in order to infer design, we need to justify the claim that goal-directedness is always the product of intelligent agency. Paley does so by offering

---

[14] This argument is similar to the reconstructions of Glass and Wolfe 1986 and Oppy 2002.

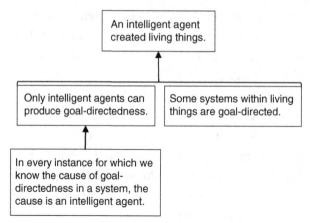

Figure 8.2   Paley's design argument interpreted as an argument from order

inductive supporting evidence. He points out that we have never seen a goal-directed system produced by one of the unintelligent processes collectively referred to as 'chance'. In other words, in every instance in which we know the cause of a goal-directed system, an intelligent agent was responsible. As Paley puts it, "In no assignable instance hath such a thing existed without intention somewhere."[15] This fully revised argument is diagramed in Figure 8.2.

It is instructive to contrast this version of Paley's argument with an inference to the best explanation. In the latter case, we have two or more competing hypotheses all of which account for the evidence. In an inference to the best explanation we attribute greater confidence to that hypothesis that does the best job of explaining the evidence, but our confidence in the conclusion is tempered by our confidence in how well we've canvassed the plausible explanations. In the argument described in this section, we begin with a premise that is like a law of physics or chemistry: all instances of goal-directedness are a product of intelligence. With this law and some facts about goal-directedness in living things we can infer deductively that a designer exists. But our key premise – the truth of the law about goal-directedness – is itself only inductively justified. So the overall argument is inductive and our confidence in the conclusion depends on how certain we are that the association between mind and goal-directedness holds. When we consider natural selection in the next

[15]   Paley 1802, 68.

chapter, we will have reason to doubt Paley's claim about the inability of unintelligent processes to generate goal-directedness. This will force us to re-evaluate Paley's argument.

In the meantime, it is worth noting an important advantage this version of Paley's argument offers over its predecessors. Specifically, it skirts Hume's most substantive objection to the argument from order (see Chapter 7). Returning to the watch example, the worry is that if one has never seen a watch made before, if one has no idea how a watchmaker could possibly assemble such a thing, then one cannot appeal to experience to justify the claim that the properties found in the watch occur only by design. Hume asserts that in order to justify the claim that a particular effect (the goal-directedness of the watch) is produced by a particular cause (an intelligent designer), one must be able to point to a great deal of experience in which intelligent designers produce precisely that effect. Paley claims that such experience is unnecessary. Nor is it necessary that we be aware of how a particular sort of intelligent agent (e.g. a watchmaker) could go about producing the effect. The implicit claim is that we can know on other grounds that only intelligence can produce this property of goal-directedness and so it matters not whether we know how, in this case, it was done. Paley points to oval picture frames and notes that, while very few of us know how they are made, we are all immediately willing to concede they are designed. Today, we might say something similar about microchips or automobiles (though I suspect most of us still don't know how oval frames are made). What goes for watches goes for organisms, and this is essential. We have no experience of an organism being created and really no idea how one would go about it. If Paley is right, then this doesn't matter – we can establish the existence of a designer without this knowledge. Of course, Paley is only right if we can establish as a general law that goal-directedness can only be produced by intelligent agency.

9   Darwin

9.1 Why dwell on Darwin and his theory of natural selection?

In this chapter we examine a scientific theory – Charles Darwin's theory of evolution by natural selection. This might seem a like a detour. After all, a scientific theory is neither a design argument, nor a direct criticism of a design argument. However, there are two reasons that an examination of Darwin's work is essential. First, Darwin explicitly casts his theory of the origin of biological species and their complex adaptations as a competitor to 'special creation', the default mode in which God was presumed to have populated the world with living things. Thus, an argument for evolution by natural selection is an argument against at least one sort of conclusion supported by design arguments. Second, Darwin's theory presents an alternative mechanism for generating the sort of order that appears in the argument from order. It thus has a direct bearing on the plausibility of one of the most popular and durable design arguments with which we've met so far.

9.2 The life of Darwin

Charles Darwin shares a birth date with Abraham Lincoln: February 12, 1809. Aside from being a curious coincidence, this fact helps to situate Darwin historically. Darwin's early life encompassed the rise and fall of Napoleon and the bloody wars that accompanied both. He witnessed the culmination of the Industrial Revolution, and from 1830 to 1846 lived through an era of electoral and governmental reform in the British Isles.[1] Darwin's thought was surely influenced by these events and by the

---

[1] Langer 1948.

profound demographic, intellectual, technological, and ideological shifts that accompanied them in nineteenth-century England.[2]

Unlike Lincoln, Darwin was born into a prominent middle-class family. His mother was the daughter of Josiah Wedgwood, his father was a wealthy physician. Darwin's grandfather was a famous physician and author of numerous works on nature that arguably presaged some of the ideas advanced by his grandson. By most accounts, including his own, Darwin was an undistinguished student. That may have had something to do with his distaste for reading the 'classics' and his inaptitude for languages. These things were all that was taught at the day-school he attended as a child. However, he took great pleasure in studying Euclid with a private tutor, and spent a great deal of time working on chemistry projects with his older brother. His chemical hobby earned him the nickname 'Gas'. Apparently, he was also quite obsessed with shooting, and developed himself into an expert marksman. Determined to prevent his son from becoming an idle child of fortune, his father sent him off at the relatively young age of 16 to study medicine at the University of Edinburgh. Darwin was profoundly bored by the curriculum, which consisted entirely of lectures. After he witnessed a surgery performed on a child (in the days before anesthetic) he aborted his medical education. As with so many of the prominent philosophers and scientists we've discussed, it was then decided that Darwin should enter the Church, and so he was sent off to Christ's College, Cambridge for three years. Again with the exception of Euclid and, significantly, William Paley's *Evidences of Christianity*, Darwin declared his Cambridge education a titanic waste of time. Of course, this isn't entirely true since it was at Cambridge that he was introduced to botany and geology and undertook his first serious scientific study. But all of this was extracurricular, as we would say. Academically, he eventually earned a BA with middling marks that failed to set him apart from the mass of his peers.[3]

What did set Darwin apart from the ordinary was a consuming interest in beetles, and this bent towards natural history would serve him well when John Henslow recommended Darwin for an appointment as 'naturalist' aboard HMS *Beagle*. The scientific ship had been commissioned to

[2] See, e.g., Cohen 1985 and Ruse 1999 for a richer historical context for Darwin's life and work.

[3] Darwin 1902.

survey the east and west coasts of South America. The vessel's five-year mission (which was in part to seek out new life if not new civilizations) lasted from 1831 through 1836, and proved to be both a transformative experience and enduring source of inspiration for the young Darwin. He first published an account of the voyage in 1839, and would later expand this account into the five-volume *Zoology of the Voyage of the Beagle*. Around this time, he also outlined an early draft of *The Origin of Species*, the seminal work which both lays out his theory of natural selection and adduces an enormous amount of evidence in its favor.[4]

However, it wasn't until 1856 that the eminent founder of modern geology, Charles Lyell, convinced Darwin to write out and publish his theory. About halfway through this enormous undertaking, Darwin received a copy of a manuscript from the young naturalist Alfred Russell Wallace that presented the identical theory, though without the massive body of data and experiment adduced by Darwin in its support.[5] Upon seeing Wallace's paper, Darwin's first impulse was to suppress his own, but Lyell and the botanist Joseph Hooker finally persuaded him to publish jointly with Wallace. So Wallace's paper along with a portion of Darwin's 1844 outline and an excerpt from an 1857 letter containing a 'short sketch' of the book Darwin was working on were read before the Linnean Society in 1858.[6] Within a year, Darwin finished his book, though it was shorter than he intended. *The Origin of Species* was first published on November 24, 1859.[7]

9.3 The structure of the *Origin*

In the first four chapters of *The Origin of Species*, Darwin provides a three-part argument. In the first part, he argues that a process he calls 'artificial selection' is effective at producing dramatically different 'varieties' or breeds. Second, he identifies the conditions that are necessary and sufficient for this process of artificial selection to occur. Finally, he argues that these same conditions are satisfied for natural populations, and therefore there must be a process of 'natural selection' producing new varieties of organisms. This basic argument structure closely mirrors the one Paley used in his *Natural Theology*. Paley appeals to the uncontroversial case of

---

[4] Darwin 1993.   [5] Darwin 1993.
[6] Cohen 1985, 288.   [7] Darwin 1993, ix.

the watch in order to identify the properties that allow us to infer design; Darwin begins with the relatively uncontroversial case of artificial selection in order to identify the conditions that must be met for new varieties to be produced. Paley finally attempts to establish the controversial conclusion of design by showing that the properties that let us spot design in the watch can be found in living things; Darwin argues for the controversial claim that organic species are produced by a process of natural selection by showing that the conditions required for varieties to be produced in artificial selection are also present in natural populations. Like Paley, Darwin also takes great pains to confront potential objections head-on. However, this is where the similarities with Paley's text end. Once the basic argument is laid out, Darwin proceeds to deduce consequences of his theory and to adduce an enormous variety of evidence bearing on each of these consequences. This part of the argument makes up the bulk of the *Origin*. The strategy is to demonstrate that the hypothesis of evolution by natural selection explains a wide range of puzzling phenomena that on the face of it have nothing to do with the origin of species. In many cases, Darwin explicitly assesses the theory of natural selection against the theory of special creation in what amounts to an argument to the best explanation.[8]

9.4 The argument through chapter 4

Darwin begins his "one long argument"[9] with a chapter on "Variation under Domestication," in which the reader is rather startlingly confronted with an analysis of the effectiveness of the domestic breeding of livestock and a discussion of the minutiae of pigeon breeding, a popular hobby amongst the upper classes in Darwin's day. In discussing the bizarre breeds of bird which fanciers have produced or the eminently useful breeds of cattle and the like, Darwin argues for a number of propositions he will need to

---

[8] Though the influence of Paley's *Natural Theology* on the composition of Darwin's *Origin* is not often raised, there are good reasons to think the similarity in structure between the books is more than coincidental. Darwin's notebooks indicate that he read Paley's *Natural Theology* sometime in the early 1840s (Darwin 1838–51), and his autobiography indicates that he held the work in great esteem, at least for its rhetorical virtues (Darwin 1902, 18).

[9] Darwin 1993, 612.

make the case for evolution of species through a process of natural selection. First, he wants it admitted into evidence that very different variants of a single species – what we might call breeds or varieties – have been produced by selectively breeding individuals from a single species, and that these variants show disparate adaptations advantageous to humans. What do I mean by adaptations? Think of the sheep-herding behavior of collies or the speed and spatial memory of carrier pigeons. While it might seem that these facts are hardly disputable in light of the enormous range of dog breeds and crop varieties that surround us today, Darwin had to make a case. Hence the emphasis on weird pigeon breeds. Darwin is able to offer compelling evidence that everything from the 'short-faced tumbler' with its comically tiny beak, to the ironically named 'runt' with its massive beak and feet all derive from a single species, the rock pigeon (*Columba livia*). He offers equally compelling evidence that in general, variants were produced by a process of cumulative selection, not through the act of preserving a single, sudden change. In fact, he cites breeding records and modern breeding practice to argue that the selections resulting in divergent modern breeds of sheep, for instance, were made on tiny, barely noticeable differences among individuals. For perhaps a more familiar example of artificial selection, just think of the enormous differences between breeds of *Canis familiaris*, the domestic dog. So far as we can tell, the panoply of breeds we know today derive from a single species, the gray wolf. It is also relevant to note that these varieties each show particular adaptations with respect to the needs or desires of humans – think again of the sheep-herding abilities of various collies and sheepdogs, the speed of the greyhound, or the unfortunate killing prowess of the pit bull. Furthermore, all of these varieties were produced by cumulative selection of small individual variations over many generations. They were not the result of a single change being captured in a pure bloodline.

Once Darwin has established that artificial selection is effective in producing varieties from a common stock, he turns to an examination of the necessary and sufficient conditions for such production to occur. In doing so, he identifies three conditions:

(1) *There must be heritable individual variation within the population.* In other words, there must be some differences amongst individuals in traits that are at least partially or probabilistically passed on to each individual's offspring.

(2) *Only some individuals are allowed to reproduce.* To put it more starkly, many or most individuals are not given the opportunity to reproduce.

(3) *Which individuals are allowed to reproduce must depend upon or be correlated with the heritable traits they possess.* It cannot be random which individuals reproduce, and at least some of the traits correlated with reproduction must be heritable.

If one of these conditions is not met, no coherent change in the breeding population is produced – no new varieties will be created. If all the conditions are met, then even widely divergent varieties of a single species can be produced.

To illustrate these conditions, think again of domestic dogs. In order to produce, say, chihuahuas, each of these conditions must have been met for a breeding population:

(1) *There must be heritable individual variation within the population.* In our dog example, it must be the case that some dogs are born smaller and that body size can be passed on to the offspring of these dogs.

(2) *Only some individuals are allowed to reproduce.* That is, only a few of our dogs are allowed to breed, namely those that are small. All the rest are never given the opportunity to reproduce.

(3) *Which individuals are allowed to reproduce must depend upon or be correlated with the heritable traits they possess.* The dogs chosen to breed are not selected at random from the population, but are by and large those that show the desired variation, namely smallness (and perhaps timidity).

Having made a case for the conditions required to sustain the uncontroversial process of artificial selection, Darwin turns in his second chapter to consider "Variation under Nature." In particular, he argues that natural populations meet the first of the three conditions given above – natural populations display enormous diversity in traits that are at least weakly heritable. Here again, examples and data accumulated from various sources describing populations around the globe are put forth in support of the straightforward claim that natural populations do in fact show variation in 'important' structures, and that at least some of this variation is heritable. To convince yourself of this fact, it's probably enough to look around at your fellow human beings. We show great variation in things like height, body shape, healthy weight, eye color, hair color, skin color, etc. and many of these features are heritable. To give a less anthropocentric example, in a typical natural population of fruit flies, hair patterns, size, eye color, wing length, etc. vary in a heritable fashion. All of this is to say that condition (1) holds for natural populations.

In discussing variation, Darwin also attempts to establish a claim that is essential for his overall argument. The distinction between varieties and species, says Darwin, is largely arbitrary and strictly a matter of degree. He argues for this claim by citing the large number of 'doubtful species'. At the time Darwin was writing, classification of organisms into species (and species into genera, genera into families, families into classes, etc.) was based on morphology, the study of form. Populations of very similar organisms were considered members of single species. A population differing significantly enough in form from a described species would be considered a distinct species. Usually, particular features or 'diagnostic traits' would be stipulated to uniquely identify a species. So, for instance, there are many features that set a moth species within the genus *Spilosoma* apart from those in other genera. But only the particular species *virginica* has orange-yellow hairs on its legs – this trait is characteristic of that species and no other.[10] However, selecting such characteristics is often rather arbitrary given the variation amongst individuals within a population. Darwin points to the fact that different taxonomists arrive at different conclusions when considering the same data. Considering the plants of Britain, one botanist (Babington) saw 251 species, while another (Bentham) classified the known specimens into only 112 species.[11] This means that Bentham considered many of the distinctions noted by Babington to represent individual variation within a species, not traits characteristic of a species. Consider one last time the great variety of domestic dogs. Looking at traits alone, one might be tempted to classify chihuahuas and St. Bernards as separate species. Another taxonomist might, given the great variation shown by individual dogs in any given population, consider these to be merely varieties of the same species. Darwin's point is that where we draw the line is arbitrary, not an objective fact about the world. Populations differ to varying degrees and when they differ a lot we tend to call them distinct species. This point is important because Darwin wants to argue that if a process can produce varieties in some amount of time, then there

[10] See www.npwrc.usgs.gov/resource/insects/macronw/6.htm (a US Geological Survey website on the macromoths of the north west). Accessed January 20, 2014.
[11] Darwin 1993, 71.

is no reason to suppose that, given more time, it could not produce species. A process that causes populations to diverge in their traits can in principle cause such a divergence that we would class these populations under new species or even genera. The point is that nothing significantly different in kind has occurred in such a process – no new 'essences' have been produced. It's just a matter of more or less change taking place.[12]

In chapter 3, "The Struggle for Existence," Darwin argues that conditions (2) and (3) above are also satisfied by natural populations. Inspired by Thomas Malthus' essay on population,[13] Darwin argues that every known species produces vastly more offspring than can possibly survive in a given area if left unchecked. That is, it is simply impossible for every organism in a population to be represented by offspring in the next generation. In natural populations, it must be the case that very few of a given generation successfully reproduce. To see why, suppose we have a tiny island on which grows only a single kind of plant – an algae, say – which is eaten by the island's only animal inhabitant – a marine iguana. Suppose that, with no iguanas at all, the island can only grow enough algae each year to feed ten iguanas. Well, if we start with a population of ten iguanas (the island is already full), and suppose fairly conservatively that each pair of iguanas (pretend there are equal numbers of both sexes) produces five viable eggs in a year, then we would have thirty-five iguanas (twenty-five new ones and ten old ones) on the island when all the eggs hatched and babies grew. But the island can only hold ten iguanas, so before the next breeding season is up, some twenty-five iguanas will have to have perished as eggs, babies, or adults.

This is Darwin's point: there is everywhere and always an intense culling of individuals from a population. Now, it is only a small step

---

[12] It is worth pointing out that taxonomy – the classification of types of organisms – changed dramatically after Darwin. Evolutionary theory provided a more or less objective approach to grouping organisms in terms of shared ancestry. I should also point out that in 1940 Ernst Mayr suggested a more rigorous definition for the basic unit of classification, the species. He said that species are "groups of actually or potentially interbreeding natural populations which are reproductively isolated from other such groups" (Curtis and Barnes 1989, 407). This definition works well for animals (though is not so apt for plants and other groups), but if embraced does not directly contradict Darwin's point. For that to be the case, one would have to show that 'reproductive' isolation cannot or does not emerge from slowly diverging traits.

[13] Malthus 1960; Cohen 1985, 293.

further to argue that the chances a given individual will leave descendants that are also able to breed depend upon the traits of that individual – the wolf with slightly longer legs is more likely to live long enough (by catching deer) to have pups than is its shorter rival. The plant that can survive two more days without water will likely leave more offspring than the one that misses the saving rain. Darwin provides more than thought experiments, of course, but we haven't the space to give details. By the end of the third chapter, then, we have established that all three conditions necessary to produce divergent varieties are present in natural populations. With all the pieces in place, it remains only to connect them.

In the fourth chapter of the *Origin*, Darwin completes the first part of his argument for natural selection: because all three conditions for the production of diverging varieties are present in natural populations, and because speciation is just the extreme divergence of varieties, we can conclude that the origin of species in natural populations is due to a selection process. Of course, there are differences between natural and artificial selection. In artificial selection, the traits that matter for determining which organism reproduces are strictly those of interest to humans. In natural selection, all traits that contribute to reproductive success in competition with other individuals and other species are favored. Also, artificial selection operates on just those variations perceptible (or easily perceptible) to humans. Natural selection operates on any and all variations, no matter how slight or inconspicuous (such as metabolic variations) as long as these variations increase the likelihood that organisms bearing them get to reproduce. This first part of the argument of the *Origin*, as laid out in the initial four chapters, is shown in the argument map of Figure 9.1. Each of the boxes containing a chapter number is a placeholder for a large, mostly inductive supporting sub-argument. Including the details would have made the diagram intractably large and obscured the central thread of argument.

Before turning to the second part of the argument of the *Origin*, I want to say something further about Darwin's strategy. It is widely acknowledged that Darwin was attempting to conform to the "best kind of science" as described by the philosophers. One of the more influential accounts of scientific reasoning to which Darwin was exposed was a slim book by

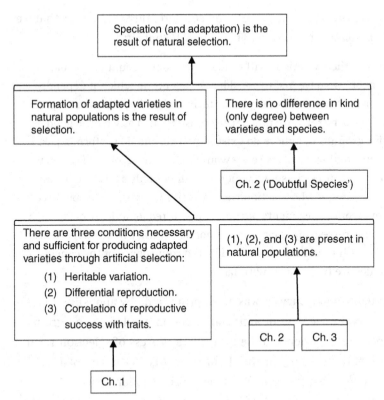

Figure 9.1   The central argument of Darwin's *The Origin of Species*

John Herschel called *A Preliminary Discourse on the Study of Natural Philosophy* (1831). Herschel, like Newton, sees scientific reasoning as a hierarchical induction. At the lowest level, we bind together particular events into universal generalizations, like Kepler's laws of planetary motion. These generalizations merely tell us what further observations to expect of a sort with which we are already familiar. Like Newton, Herschel refers to these low-level generalizations as phenomena. So, for instance, the generalization that the planets travel in ellipses is a phenomenon. Scientific theories aim at explaining phenomena, generally of more than one kind. For an example, think of Newton's theory of universal gravitation. This tells us why the planets travel in ellipses and also entails a great many more kinds of phenomena such as Kepler's other two laws of motion, the existence of tides, etc. What Darwin is offering us in the *Origin* is a theory – an account that explains a wide range of biological phenomena. But how do we justify a hypothetical theory as the correct choice? Here, Herschel insists that we choose a *vera causa*, a true cause. This is how he puts it:[14]

> In framing a theory which shall render a rational account of any natural phenomenon, we have *first* to consider the agents on which it depends, or the causes to which we regard it as ultimately referable. These agents are not to be arbitrarily assumed; they must be such as we have good indicative grounds to believe do exist in nature, and do perform a part in phenomena analogous to those we would render an account of ... They must be [true causes], in short, which we can not only show to exist and to act, but the laws of whose action we can derive independently, by direct induction, from experiments purposely instituted; or at least make such suppositions respecting them as shall not be contrary to our experience, and which will remain to be verified by the coincidence of the conclusions we shall deduce from them, with facts.

This is arguably what Darwin was attempting to do by appealing to artificial selection. In the case of artificial selection, we have good grounds from immediate experience to believe that a process of selection results in divergent varieties of organisms. In other words, in the case of artificial selection, we are able to identify a true cause of biodiversity. When we turn to natural populations, all of the same conditions are met as for selection in the artificial case, and so we can infer that, given the similarity of the effects, the same true cause is responsible. Of course, there are some differences in the two cases. In natural selection, no one is consciously choosing which organisms will reproduce. To hedge his bets, Darwin will also attempt to verify his supposition "by the coincidence of the conclusions" he can deduce from it.

9.5 Consilience and the consequences of natural selection

If the argument with which Darwin begins the *Origin* is constructed with Herschel's view of scientific induction in mind, the arguments with which he finishes it bears the mark of William Whewell. In Chapter 6 of this book, I briefly discussed Whewell's theory of scientific method. Roughly, Whewell thinks that inductions are confirmed by consilience. That is, a

[14] Herschel 1831, 197.

hypothesis is confirmed when it explains facts in a wide variety of kinds of phenomena and particularly when it explains phenomena that were not considered when forming the hypothesis. This is similar to a simpler model of scientific inference now called 'hypothetico-deductivism'. In this model, one advances a hypothesis, deduces consequences of the hypothesis, and then checks these against empirical data. This is essentially the pattern of the remainder of the *Origin*. In chapter 4 (and throughout the rest of his book), Darwin proceeds to examine some of the consequences of the mechanism he is proposing. Natural selection, if it occurs, would entail some facts about the world besides the adaptedness of organisms. In many cases, Darwin also contrasts natural selection with special creation – the repeated creation of species by God throughout geological time – and argues that the former is a better explanation. Thus, much of Darwin's argument consists in showing that not only does evolution by natural selection explain an enormous range of phenomena beyond the mere origin of species, it does a better job in each case than special creation. In this way, many inferences to the best explanation combine to satisfy Whewell's criterion for confirming natural selection.

I wish we could discuss all of Darwin's examples, but we have space for only three of them. To begin with, Darwin notes some curious patterns amongst species. For instance, those species which demonstrate the greatest individual variation tend to belong to genera with lots of different species.[15] Darwin supported this conclusion by counting species and varieties reported in taxonomical indices of plants and beetles.[16] In the *Origin*, Darwin doesn't present any specific numbers. To give an example, I'll point to the beetles but I'll use family, genera, and species rather than genera, species, and varieties to make the point. Within the order *Coleoptera* (the beetles), the family *Aldiridae* (ant-like leaf beetles) there are 7 genera containing an average of 1.6 species. In the family *Carabidae* (the ground beetles) there are 104 genera with an average of 3.7 species in each.[17] Now obviously, this is one example chosen for illustration. But

[15] Darwin 1993, 81–82.

[16] For a detailed account of how Darwin gathered and analyzed the data supporting the claims in this section of the *Origin*, see Parshall 1982.

[17] These numbers are the result of calculations by the author based on the catalogue of species available at: http://bugguide.net/node/view/60/tree, accessed February 25, 2009. The counts of species do *not* include subspecies.

the pattern holds – the more diverse any given taxon is at one level, the more diverse it tends to be at the next higher level. Why should this be the case? If natural selection really is responsible for the origin of species (and genera, etc.), then the observed pattern is not surprising. Populations with lots of individual variation permit the creation of diverging varieties at a high rate. As varieties diverge from another, they produce species. As clusters of species diverge from another, they produce genera, and so on. If natural selection is the process that produces species, then variation at one level should lead us to expect variation at the next higher order of classification, and vice versa.

By the time Darwin was compiling material for the *Origin*, the new science of paleontology had already produced some significant results. Fossils of animals that no longer exist anywhere in the world had been known for some time, but patterns had emerged relating the extinct forms to extant forms of life. I'll give a modern example that is particularly striking. The primates are a very recognizable group of mammals that includes lemurs, monkeys, and apes. The fossil record of the last 60 million years boasts a number of extinct species that bear more or less strong resemblances to modern primates. That is, various extinct species possess some or all of the characteristics diagnostic of the group of primates. This is also the case for the lagomorphs – the rabbits and picas. But there is a curious pattern amongst these extinct species: as we look back in time, we find only species of intermediate character, species with some features similar to the characteristic traits of primates and some similar to the characteristic traits of lagomorphs, with nothing that quite looks like a modern ape or rabbit. A cartoon diagram of the situation appears in Figure 9.2. In that figure, each black dot represents a known fossil of an extinct species. The vertical axis indicates how old the fossil is, while the horizontal axis is supposed to represent morphology or form. As you move right on this axis, you describe animals that look more and more like rabbits, as you move left you describe animals that look like primates. Note that the further back you go, the more fossil species are confined to the intermediate region. Now, Figure 9.2 is just a cartoon – the dots don't correspond exactly to the known fossil record. But Figure 9.3 depicts the skulls of a modern gorilla and some known fossil primates that bear increasing resemblance to lagomorphs as we move back in time.

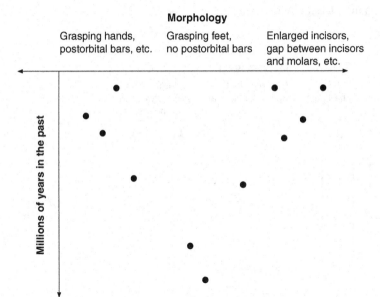

Morphology

Grasping hands, postorbital bars, etc.   Grasping feet, no postorbital bars   Enlarged incisors, gap between incisors and molars, etc.

Millions of years in the past

Figure 9.2   A cartoon representation of the fossil record pertaining to the primates and lagomorphs. Each dot represents a known fossil of an extinct species. The vertical line indicates the age of the fossil (the lower, the older) while the horizontal axis represents biological form. Moving to the left means possessing a greater number of traits characteristic of apes, monkeys and lemurs, and to the right a greater number of traits characteristic of rabbits

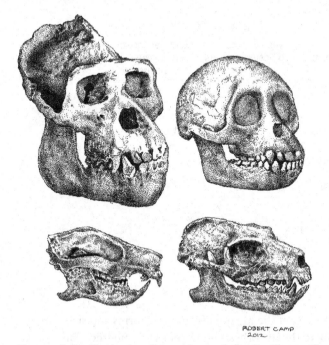

ROBERT CAMP
2012

Figure 9.3   Clockwise from upper left: *Gorilla gorilla* (modern), *Proconsul sp.* (25–23 Ma), *Notharctus sp.* (57.8–36.6 Ma), and *Plesiadapis sp.* (66–57.8 Ma). Illustrations by Robert Camp

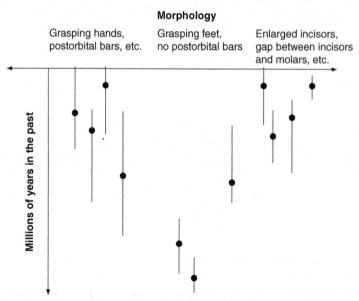

Figure 9.4   Reconstruction of the history of biological forms appearing in the fossil record according to the hypothesis of special creation

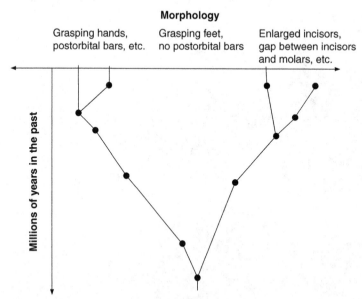

Figure 9.5   Reconstruction of the history of biological forms appearing in the fossil record according to the hypothesis of descent with modification

Why is this pattern significant? Well, there are (at least) two ways to understand this pattern of change. Under the hypothesis of the special creation of each species by an intelligent designer – such that species only arise by special creation – we would have to construct the diagram as shown in Figure 9.4.

Species come and species go, but do not change. However, in this view it is only a strange fact that intermediate species were created and destroyed in the past and only diverging species are extant today. On the other hand, if we allow that species come from species, we might fill in the diagram as shown in Figure 9.5.

This is just what Darwin did in the only diagram to appear in the *Origin*, and it is just what one would expect if natural selection were responsible for producing species. In short, this is because only by diverging in charac-ter can closely related species avoid direct competition with one another and exploit novel ways of making a living. Darwin dedicates a great deal of chapter 4 to analyzing the expected process of species development.

Finally, Darwin also considers some facts about biogeography, the dis-tribution of species in space. It is interesting, for instance, that islands gen-erally have a relatively low biodiversity but a high proportion of endemic species, species that are found on that island and nowhere else. This is difficult to account for on the hypothesis of special creation. To be more clear, this fact is *not incompatible* with special creation, but there would be no reason to expect it on grounds of intelligent design. Why should a designer (an entity with a mind and motive somewhat like ours) choose this pattern over any other (such as a uniform number of species in each environment)? On the one hand, were natural selection responsible, then these facts would be unsurprising. On a small landmass, there is little competition amongst the one species that manage to colonize the island – there are lots of open ecological niches. Thus, one would expect a low rate of species production. On the other hand, there is no regular mixing with individuals from off the island, so one would expect the populations on the island to diverge from those on the mainland.

9.6 Darwin and design arguments

So what does Darwin's theory of evolution by natural selection have to do with design arguments? In short, if Darwin is right about where species come from and how it is that organisms come to possess structures adapted to a particular environment, then it becomes much harder to make a compelling argument from order. So far, the most plausible version of the argument from order we've considered was a refinement of Paley's argument as presented in the preceding chapter. That argument takes it for granted that a property I called 'goal-directedness' can only be produced by an intelligent agent. But Darwin's theory offers an alternative explanation for how such goal-directedness is developed. The eye that sees with great acuity, the heart that pumps blood, the wings that allow a condor to glide with great efficiency: all of these are, according to Darwin's account, the products of a process of natural selection. Put more starkly, Darwin argues that the features of living things previously thought to indicate that they were crafted by some intelligence to satisfy a purpose are in fact the outcome of another causal process. The goal-directedness of organisms is not illusory or coincidental, in this view. Rather, it is the outcome of a sort of causal process that was previously unconsidered. If that's true, then we deprive the argument from order of its most important premise; there are other sources of goal-directedness than intelligence, and so merely detecting the former does not allow us to infer the latter.

There are, of course, a variety of responses open to the design proponent. First, one could attempt to argue that Darwin's theory is wrong. This is not a promising tactic in light of the success of modern evolutionary theory and the mountain of evidence that supports it. Alternatively, one could refine the relevant notion of order. Perhaps goal-directedness is not quite the right property to focus on. Perhaps there is a related property that is plausibly linked to intelligence in the right way. Of course, it's unlikely to be a property that organisms have, since we seem to have a satisfying theory of the proximal cause of the properties of organisms. But maybe this is a property displayed by the process of evolution itself. Or perhaps it is a property possessed by some other kind of thing in the universe. Without a specific proposal on the table, it's hard to say. In the second half of this book, we'll consider a few modern proposals and see if they fare better than Paley's. I should stress that Darwin did not expose some sort of error in the form of the argument from order. Rather, his theory undermines a key premise. It remains to be seen whether a suitable substitute can be found.

10   Loose ends

10.1 Overview

In our survey of the first twenty centuries or so of design arguments, we covered a lot of ground. In the process, many issues worth considering were raised but quickly set aside. This chapter collects a number of these issues together and gives each due consideration. The reader in a hurry can skip this motley collection without losing the thread of debate. But for those with nagging questions about the idea of purpose in nature, how much natural theology can establish about the attributes of God, or the connection between William Paley and Bernard Nieuwentyt, what follows will be of interest.

10.2 Purpose in nature

In Cicero's dialogue *De natura deorum*, there appears an argument that rests on the observation of 'purpose' in the universe. This argument resurfaced centuries later in the great compendia of natural history such as Derham's *Physico-Theology*. Despite the longevity of this sort of argument, we quickly passed over it with little scrutiny when it came up in our survey. It's time we rectified that oversight.

Arguments from purpose begin by pointing to natural things which clearly serve a purpose. Typically, these are the adaptations of organisms. For example, the extraordinary ovipositor of the parasitic wasp *Megarhyssa* obviously fulfills the function of drilling through wood, and Derham marveled at the ability of the gall wasp's eggs to alter the development of plant tissues so that the plant constructs the perfect home for the gall wasp larvae. More familiarly, the human eye is exquisitely adapted to the task of seeing and the human hand for grasping. We could list enough examples from the living world to fill numerous books this size. In fact, many authors have done so over the last few centuries. But to make the argument we need only one clear example of purpose in nature. The rest of the argument is trivial. Only an intelligent agent can arrange things for a purpose. After all, that's what it means for something to have a purpose – some intelligent agent wills it to fulfill a particular function. Thus, there exists a designer of at least one natural object.

A problem immediately presents itself with the argument as I've just put it. If what we mean by saying that some natural structure like the eye has a 'purpose' is just that it was designed by some intelligent agent, then we argue in a circle when we take it as a premise that the eye has a purpose! To put it another way, all of the observations presented as empirical evidence for purpose in nature – e.g. that the eye has the purpose of vision – really amount to statements of the conclusion we are supposed to be arguing for. Someone who doesn't already accept the conclusion would be rationally justified in simply denying these premises. As presented, the argument from purpose is no argument at all.

One way we might avoid begging the question is to try and identify a set of properties that do not refer to the will or goals of an intelligent agent but that nonetheless make a thing seem to have a purpose. In other words, we can try to explain what kind of thing looks as if it has a purpose, but do so without assuming that such a thing really was produced by an intelligent designer. If we're lucky, we'll be able to show that the things that look as if they have a purpose really do have one. This is similar to the recipe for building an argument from order, though starting from a slightly different intuition. And this approach has been tried. One notable example from the late nineteenth century appears in a book by Paul Janet.[1] Janet noted the question-begging difficulty highlighted above, and in response he separated the question of purpose in nature into two questions. First, what sorts of things in nature evoke an intuition of purpose, and second, what does the existence of such things imply about the existence of a creative intelligence? Here is the answer Janet proposes to the first question:[2]

> When a certain coincidence of phenomena is remarked constantly, it does not suffice to attach each phenomenon in particular to its antecedent causes; it is necessary also to give a precise reason for the coincidence

[1] Janet 1884.   [2] Janet 1884, 60–61.

itself ... When a certain coincidence of phenomena is determined, not only by its relation to the past, but also by its relation to the future, we will not have done justice to the principle of causality if, in supposing a cause for this coincidence, we neglect to explain, besides, its precise relation to the future phenomenon.

In other words, there are some phenomena – for instance, the vertebrate eye – that involve the coming together of a great many causal chains in a very precise way. Were any of the processes of embryonic development disturbed in rather minor ways, no eye capable of vision would be formed. Furthermore, phenomena like the eye are robust in that they keep occurring in the same way, over and over. So far, this sounds very much like the 'marks of design' examined by Boyle, Nieuwentyt, Paley, and others. However, according to Janet, phenomena that evoke a sense of purpose have an additional feature not yet considered: the details of each of the many causal processes that must coordinate to produce one of these phenomena are insufficient to explain why these causal processes are constantly joined together. It takes a lot to make an eye that can see with the exquisite clarity of the human eye. Lots of processes must come together just right: corneal cells must arrange themselves into an appropriately curved and transparent layer, a crystalline lens must be formed by surrounding tissue in just the right shape and be placed in just the right location, and so on. If we pay attention to the molecular details, the complexity of the overlapping causal processes is overwhelming. But tracing any one of these processes (or all of them individually) gets us no closer to explaining why they reliably and repeatedly coordinate to make an eye. Janet's claim is that there is nothing in molecular physics or even cell biology that tells us to expect the repeated formation of an eye. We might expect cells to divide or crystals to form, but there is no reason to suppose these processes to constantly combine. The only way to explain that, says Janet, is to note that the end result is an eye that sees. The co-occurrence of all of these processes is explained by the nature of the end result they will produce. In other words, when asked why so many cellular and molecular processes keep occurring together and in such an intricate pattern, the only plausible answer is that they are directed toward the production of an eye that can see.

Granting Janet this much of the argument, we can ask the second question: what can we infer from this regarding the existence of a deity?

Supposing we must refer to the future end state of a certain complex collection of processes in order to explain why that collection keeps occurring, does this mean that the processes must be directed by an intelligence toward the end state we've identified? Janet thinks the answer is yes. Certainly, we may concede that in many cases involving the intentions of an intelligent agent, the resulting phenomenon has the properties Janet indicates. For instance, it takes quite a few intricately balanced causal processes to produce a blender. The only reason these processes ever occur together is because people want a device with the properties of a blender, and so they bend nature to their will in order to produce one. But the relationship is not one-to-one. Not everything intentionally created for a purpose is complex; a reservoir made by damming a stream with earth is about as simple as it gets. Nonetheless, it has a purpose, and a very important one at that. So the kind of complexity Janet points to is not necessary for a thing to have a purpose. But is it sufficient? To show that it is not, we need only demonstrate that it is possible for something to exhibit the "coincidence of phenomena" that Janet has in mind and yet fail to be the product of intelligent agency. A wide variety of counterexamples suggest themselves. While it takes an enormously complex collection of causes to produce a footprint, for instance, no one thinks footprints have a purpose. The same applies to most of the by-products of life. To take a cheeky quotation from John Stuart Mill slightly out of context, "no one sees marks of design in a coprolite."[3] Yet fossil dung (yes, that's what a coprolite is) requires an even more extensive confluence of causes to produce it than does the eye. Of course, these are isolated and somewhat peculiar examples. And they all involve effects of a kind of thing – living organisms – whose design status is in question. One might suspect that we could fix our criteria so that the organism is ruled in and extraneous effects like footprints and fossil excrement are ruled out.

But there is a more systematic problem with Janet's proposal. We can show that all of the examples from the biological world that are typically invoked in arguments from purpose can be explained without intelligent causes. These examples include essentially any complex trait of an organism where it is clear how that trait contributes to the organism's survival and reproduction. For these traits, evolution by natural selection offers a

---

[3] Mill 1998, 168.

powerful explanation. The reason organisms exhibit the coordinated casual processes that result in a trait like the eye is that their ancestors had
this capacity and were selected because of it. That is, those systems (organisms) with the combination of causes that result in an eye fared better and
reproduced more than organisms with different sets of causal dispositions.
According to the evolutionary view, Janet is right in a sense – we can only
explain why we find in organisms the very complicated suite of causal processes required to develop an eye by appealing to the final product, namely
the eye and the sense of vision it provides. But the evolutionary appeal to
the eye is backward-looking. It is because eyes were selected in ancestors of
living organisms that current organisms have the capacity to produce eyes.
So evolutionary theory offers an explanation of traits with all the properties identified by Janet, but does not invoke an intelligent designer. Thus,
apparent purpose in Janet's sense does not imply purpose in the intentional
sense. Traits with a natural purpose do not entail a designer.

There is another strategy we might pursue in order to avoid begging the
question in an argument from purpose. This strategy is much more radical.
To see how it would work, we should turn to a classical source: Aristotle.
Aristotle provided the first comprehensive theory of causation, where by
'causation' I mean something much broader than is implied by the modern
use of the term.[4] Roughly, to ask for the cause of a thing or event in the
sense Aristotle intends is to ask for an explanation. As you might imagine,
there are more ways to explain than identifying what we would call causes
today. Aristotle classified the kinds of explanation or 'causes' into four categories. One of these, the *efficient cause*, is closest to the modern notion –
it's the immediate source of change, the immediate reason something has
come to be. To cite one of Aristotle's own examples, an efficient cause of the
child is the father (and, presumably, an activity of the father's). The other
types of cause are *material* (what a thing is made out of), *formal* (an account
of what it is to be that sort of thing), and *final*. The term 'final cause' is a
product of medieval scholasticism; Aristotle referred to such a cause as
a τέλος (or *telos* when anglicized), which just means 'end' or 'purpose'.[5] A
final cause is that for which a thing exists or a process occurs; it's the purpose that thing or process is supposed to satisfy. There is something very

---

[4] For a general overview of Aristotle's theory of causation, see Falcon 2012.
[5] Janet 1884, 1.

important to note about Aristotle's use of the term 'purpose'. He is explicitly attempting to offer an account of natural phenomena that steers a middle course between what he sees as two defective approaches: the strict materialism of the Ionian natural philosophers which leaves no room for purposes, and the theological account of Plato in which the world is an artifact produced by a divine craftsman and in which everything has a purpose. So when Aristotle refers to 'ends', he does not mean it in the intentional sense.[6] Rather, ends are unintentional features of a thing or process; they are inherent principles in things which lead them toward a particular final state that has no direct dependence on the will or the intentions of any mind.[7] This is somewhat ironic given how later philosophers made use of his theory. Once Aristotle's metaphysics was embraced by Christianity, the notion of final cause was immediately annexed to the idea of God's will. In the hands of the scholastics, final causes were precisely what Boyle would later take them to be: God's purposes. In fact, it is from the Greek word *telos* that we get the English word teleological, meaning goal- or purpose-driven. One often hears the argument from purpose referred to as the 'teleological argument'. But we need not follow the scholastics down that road. Instead, we can stick with something closer to the original. Now it is true that Aristotle attached moral value to the final states – he asserts that the ends toward which things naturally tend are good. But there is no reason we can't jettison this part of the theory while retaining the essential framework of explanation by appeal to natural goals.

Arguably, this is precisely what we have in modern physics in the form of 'variational principles'. These are physical principles which assert that some special function of the physical quantities must take that form which is minimal or maximal. The history of variational principles begins in the mid eighteenth century with the publication of Pierre Maupertuis' Principle of Least Action. This principle states that nature always minimizes a quantity called the 'action'. For Maupertuis, the action is the speed of an object times the distance traveled over each segment of its path, summed over its entire path for a given interval of time.[8] Though

---

[6] See, e.g., Lennox 2001, Part III.   [7] Ward 2003.

[8] For a history of Maupertuis' discovery of the Principle of Least Action, see Fee 1941 and Jourdain 1913. Interestingly, it was a concern with teleology – the question of why God would choose an inverse-square law of attraction rather than some other – that led Maupertuis to the Principle.

Maupertuis overstated the scope of his principle (it works only for certain classes of physical phenomena), it was nonetheless striking for its generality. Here's how Maupertuis himself describes it:[9]

> There is a principle truly universal, from which are derived the laws which control the movement of elastic and inelastic bodies, light, and all corporeal substances; it is that in all the changes which occur in the universe … that which is called the quantity 'action' is always the least possible amount.

By the end of the nineteenth century, three great mathematicians, Euler, Laplace, and Hamilton, had developed Maupertuis' vague and limited principle into a profoundly general framework in terms of which all of classical mechanics (and more) was recast. This new formulation of the physics rests upon Hamilton's Principle:[10]

> Of all the possible paths along which a dynamical system may move from one point to another within a specified time interval … the actual path followed is that which minimizes the time integral of the difference between the kinetic and potential energies.

This descendant of the Principle of Least Action has played a pivotal role in the development of modern quantum theories – it is by no means a dusty relic of the eighteenth century. What matters for our purposes is that both of these variational principles are instances of Aristotelian final causes. All physical systems – in fact the physical universe as a whole – tend toward the end or goal of minimal action. In other words, we have in physics a profound example of a natural purpose along the lines Aristotle himself proposed. In this case, the minimization of this curious quantity 'action' is the end toward which all physical systems tend. We can frame nearly our entire science of physics in terms of this final cause. So it is evidently possible to frame the notion of purpose in such a way that we do not beg the question in favor of design when we attribute purpose to things like the vertebrate eye. By adopting the Aristotelian view, we have removed the principal impediment to a successful argument from purpose. In the process, we have gained an interesting perspective from which to do natural science. However, we have lost the easy road to a deity. Precisely because Aristotle's account allows us to conceive of a great many phenomena as having a

[9] Fee 1941, 503.   [10] Marion and Thornton 1995.

purpose – including things that we're confident are not the direct products of intelligent agency – we cannot simply infer design from purpose. Nor is it obvious what else would be needed – what features a thing must have in addition to natural purpose – in order to make such an inference.

10.3 How many gods?

We have spent a great deal of time examining inferences from observable features of the world to the existence of deity. Typically, such an inference begins by pointing to some phenomenon in the universe – e.g. the motions of planets, the structure of human bodies – and then argues that each instance of such a thing has one or more features that suggest it is the product of design. If successful, such an argument would directly establish that there exists at least one designer. But it is logically possible that every single item from which we can argue to design has its own designer. The Stoics embraced just such a conclusion, positing a distinct deity for each of the heavenly bodies that exhibit regular motion. Aristotle's view was not too different. Even in the eighteenth century, we find Hume arguing that the existence of many lesser creators is a more plausible conclusion from the evidence than a single designer of immense power. Yet for most, a successful design argument is supposed to establish the existence of a unique designer – the existence of a single God.

How might we establish the unity of the designer given the great diversity of things that are designed? Granting that any one of the arguments we have considered so far is sound, at least as regards the inference from empirical evidence to the existence of at least one designer, how might one further demonstrate that there is just one responsible for all? There are two basic strategies one can pursue. One appeals again to features of the world and argues for the unity of God from the evidence at hand. The other approach appeals to metaphysical principles that are independent of the question of design. William Paley employs a strategy of the first sort in chapter 25 of his *Natural Theology*. After demonstrating that organisms are products of design, he argues that each exhibits a common plan. That is, there is a great similarity in the design of each organism, so much so that we are entitled to infer a single intelligence behind them all. What similarity? Well, amongst the vertebrates, we find all the same tissue and organ types: blood, bone, skin, heart, liver, etc. What's more, even the

casual student of comparative anatomy must be struck with the similarities amongst vertebrate skeletons. It is possible to map (almost) all of the
bones in my body to corresponding bones in the body of a deer or elephant, with each in the same relative position. A dog has the same bones
arranged in the same sequence in its leg as I do, though they are shaped
differently (and the dog has one less toe). As we expand our view to encompass more and more of the animal kingdom, we can continue to trace such
structural similarities. The human body plan is not so different from, say,
that of a fish. However, the human body plan is enormously different from
that of an insect or a starfish, and it is here that Paley's argument begins
to lose its intuitive force. As long as we stick with the obvious affinities
between our bodies and those of other mammals, it doesn't seem necessary to demand an account of how similar things must be in order to have
the same plan or design. However, Paley wants to say that virtually all
living things exhibit the same basic plan. It is not at all obvious that I and
a grasshopper are really variations on the same design, so the need for an
account of sameness is pressing if we are to make such a claim convincing.
Unfortunately, it is difficult to see how to give such an account without
trivializing the notion of sameness. We would have to find some principled way to avoid conceding that every conceivable living thing has the
same plan. Paley struggled with this problem as he stretched to include
the insects amongst the living things with basically the same plan of construction or design as we humans. He claimed that insects and shellfish
are of an analogous plan to vertebrates because they exhibit the inver-
sion of the vertebrate body plan. Insects and shellfish have their hard,
rigid parts on the outside rather than on the inside as a skeleton. This fact
supposedly "confesses an imitation, a remembrance, a carrying on of the
same plan"[11] as the other animals. But by these lights, there is no difference that could possibly represent a distinct plan. Consider an animal
with no hard parts – a jellyfish, say. We could always claim that its absence
of hard parts reminds one of the missing skeleton of a vertebrate, and is
thus a carrying on of the same plan. Clearly, if this strategy is to yield a
compelling argument for the unity of the creator, we have to know how
to assess the sameness of design in a way that does not beg the question
about the number of designers. Of course, it is almost certainly a vague

[11] Paley 1802, 486.

question as to where precisely the boundary is between distinct designs. But at the very least, we need a clear criterion for declaring two things to be of different design – just how different must one animal be from the others before it unequivocally counts as a distinct design? No such account has been given, though that does not mean it is impossible to give one.

Turning to the second and more promising strategy, one might appeal to Occam's razor. Occam's razor is a metaphysical principle named for the fourteenth-century philosopher William of Occam, though in its modern forms the principle is only loosely connected with his philosophy. As it is used today, it's a pithy principle about theoretical simplicity:[12]

(OR): Entities are not to be multiplied beyond necessity.

The idea is that we should always favor the simplest theory, that is, the theory that invokes the fewest things. It should be obvious how this little principle can help us with our design argument. Once we've established that there exists at least one designer, we can just appeal to OR to conclude that there is only one. After all, one God is a simpler theory than many gods. Or is it? The pithiness of OR comes at a price, namely ambiguity. There are two ways we can read the principle. We might see it as urging us to adopt the fewest number of individual things we can. If we can make do by supposing that one thief was responsible for the heist, we shouldn't hypothesize more than one. If we can account for the motions of the visible planets by the hypothesis that there are three more, then we shouldn't suppose that there are five. Under this reading, we get what we want – if one God will do the job, we shouldn't postulate more.

But we might instead see OR as urging us to posit the fewest *kinds* of things possible. If we make do with ninety-eight chemical elements, we shouldn't postulate two hundred. If material alone will suffice, we shouldn't posit mental substance in addition. Under this reading of OR, it looks as if we should favor the hypothesis of many designers of finite capacity. That is, we should suppose that there are many intelligent agents like us rather than one very different sort of deity. Of course, you might object that any designer of, say, the human eye must be very different from us and so we have already been forced to cross the threshold and posit a new kind of entity. But this only draws out an underlying problem. Even

---

[12] Baker 2011.

if we can offer a clear defense of this version of OR, how it applies to any given case depends crucially on how we divide the world into kinds. This question is far from settled. In our case, the specific question is whether we should count a cosmic designer of this or that ability as a different kind of thing from the intelligent agents with which we are already familiar (e.g. people).

So far it looks as if we should insist on the quantitative version of OR. This doesn't seem such a bad choice, since the principle is at least intuitively plausible (there are interesting defenses of this principle). But if we adopt the quantitative version of OR, we face a new problem. The question is what collection of things we're minimizing. OR is most plausible given assumptions about what sorts of things exist. Once we know we're postulating unseen planets, then we should postulate as few as possible. But we first have to have a reason to postulate unseen planets. In this case, we have to have a reason to postulate omnipotent (or, at least, ineffably powerful) designers. Once we've done that, OR lets us conclude that there is but one of them. However, it is not at all clear that the evidence supports this hypothesis. That's what Hume was getting at. If in each case of design, the evidence favors a finite or limited designer, then OR tells us to postulate as few as we can in order to account for all of the instances. But this number is likely to be greater than one. So it seems that in order to invoke OR to settle the question of unity, we must first settle the question of the designer's attributes. That is the subject of the next section.

10.4 What's God like?

Again, suppose we have a sound argument that leads us to conclude that some portion of the universe – or perhaps the universe itself – is the product of intelligent agency. Then we know that there exists at least one creative intelligence responsible for the existence or properties of that part of the cosmos. What can we say about the attributes of that designer? Can we prove it likely to be an omnipotent and loving deity, or are we only able to establish something less? Various authors we've encountered have taken up these questions. Paley, for instance, devotes the end of his *Natural Theology* to demonstrations of God's goodness and other "natural attributes." John Stuart Mill dedicates Part II of his essay on "Theism" to a consideration of the attributes of God that can be inferred from observable evidence.[13] Much of what I have to say in this section owes a debt to Mill, but it is not my intention to examine any particular treatment of the problem. Rather, my aim is to give you a sense of how far one might get and where the difficulties lie.

I should first clarify the rules of the game. Supposing we have a sound argument for the existence of a designer, we are attempting to figure out what attributes the deity has using only the same sort of empirical evidence with which we started. The goal is to give an a posteriori argument from facts about the world to characteristics of the designer. This is a very different goal from trying to show that the world as we find it is compatible with a God that has certain attributes like omnipotence and omniscience. While the world might be compatible with such things, it might not offer any evidence that they are true. Let me illustrate the difference, because the distinction is often lost in discussions of natural theology. Think about the Antikythera Mechanism, the extraordinary piece of ancient clockwork pulled from the bottom of the Aegean (see Chapter 1). We know how old the object is (about 2,000 years). And we're pretty sure it's an artifact – a product of intentional design. We were quite surprised to discover such a thing of that age because there is no record of comparable machinery until the Middle Ages. So where did it come from and what was it for? The archeologists and other scientists working on this problem are attempting to determine what we can rationally infer with some confidence from the available evidence. They want to answer the question, "Given the evidence we have, what are we entitled to conclude about the origin and use of this device?" Now someone else might already – for reasons independent of the evidence surrounding the Antikythera Mechanism – have a theory that bears on the origin of that device. For instance, someone interested in making a sensational program for a cable TV network (I wish this were a fanciful scenario) might insist that extraterrestrials visited the ancient Greeks and gave them many high-tech gifts, including the Antikythera Mechanism. Such a person would not be interested in arguing from the evidence to this conclusion, especially since the evidence won't get you there. Instead, she is likely to argue that the space alien theory is compatible with the available evidence. That is, there is nothing known about the

---

[13] Mill 1998.

Antikythera Mechanism that contradicts the extraterrestrial theory. For the purposes of this book, we have no pre-established thesis to defend. We are interested in going wherever the empirical evidence takes us. So the problem at hand is like that of the archeologists: what can we infer about the designer's attributes from facts about the way the world is? We are not interested in demonstrating that the facts are compatible with one or another theological position.

Let's begin with the moral attributes of the designer, and in particular the question of God's goodness. Our evaluation of the evidence will depend on what we may infer as the likely goals of the designer, and this in turn depends on precisely which parts of the world we take to be designed. But we'll capture a very wide range of cases of interest if we focus our attention on living things. Here, the evidence is quite mixed. On the one hand, as many authors in natural theology have pointed out, animals are equipped with a sense of pleasure that seems to exceed that which is necessary for survival. On the other hand, we are also equipped with a sense of pain and ample opportunity to use it. On the one hand, our bodies are exquisitely well adapted to provide for their own perpetuation. We have ample tools (our hands and big brains) for gathering food, securing shelter, finding mates, and so on. The same can be said for every other living thing – each is exquisitely adapted for its own survival. On the other hand, every living thing is adapted for its own survival at the expense of some other living thing. From the perspective of the larva that gets eaten alive when the injected egg develops, the exquisite ovipositor of the *Megarhyssa* wasp is a ghastly implement of torture. For the dove, the talons of the falcon are hardly for the best. When it comes to living things, then, the only apparent goal of their design is to perpetuate the system of living things – each type of organism is designed to perpetuate that type, and the set of organisms coordinates in such a way as to more or less perpetuate the set. This is all carried out with both pleasure and a great deal of pain. While the capacity for pleasure may be taken to demonstrate some concern with the well-being of living things, the evidence does not suggest that the world is arranged to give each organism the greatest possible amount of pleasure. Much the same can be said about any moral good one cares to identify. The world presents a mixture of beauty and ugliness, for instance. For some values, the evidence paints a worse picture. Mill takes the rather pessimistic view that there "is no evidence whatever in Nature for divine justice, whatever standard of justice our ethical opinions may lead us to recognize."[14] Both the best and the worst of us are afflicted with diseases, accidents, and other natural evils. The wicked profit as much as the good from whatever bounty nature provides. There is no meting out of justice by the natural world, and so no evidence for a just designer. What can we conclude from these considerations? Only that the designer has some concern for the good of living things. This is a far cry from perfect goodness or boundless love. Let me stress again that I am not suggesting that the evidence precludes the designer from having the attributes of perfect goodness and boundless love, only that we cannot infer these things from the empirical evidence alone. Such a conclusion seems beyond the scope of a posteriori natural theology.

The prospects for justifying belief in a good or just designer on the basis of the empirical evidence seem poor. But what about omniscience and omnipotence, two attributes generally attributed to the God of the Abrahamic religions? If we are speaking merely of the creator of this or that living thing or even all living things, then the designer must have a vast but finite knowledge and a corresponding ability to manipulate the physical world. But such capacities are, well, infinitely far from unlimited power and total knowledge. In fact, it isn't clear how far from our own capacities such a designer must be. If we expand the scope of our design argument so as to appeal (successfully, for the sake of argument) to grander portions of the universe like the solar system, then our estimate of the designer's capabilities must proportionally increase. But as long as we are speaking of finite accomplishments, we will only be able to infer limited knowledge and power.

In fact, Mill argues that so long as our argument is based on the detection of design or contrivance – the adjustment of means toward ends – then the designer cannot be said to be omnipotent on pain of contradiction. This is because an omnipotent being has no need of means or of contrivance. An omnipotent being can simply will whatever state of affairs she wishes. There are only ends for such a being; she has no need of means. Design on the other hand presupposes an agent operating with constraints, configuring what she can in the world so as to achieve a particular end she cannot achieve otherwise. This objection carries force only insofar as we think our design arguments establish 'design' in the sense Mill has in mind. It

---

[14] Mill 1998, 194.

is true that one cannot consistently assert that the universe shows signs of contrivance and still insist that the creator of the universe is omnipotent. However, some arguments we have considered conclude only that an intelligent agent has intentionally played a role in the outcome we have before us, not that the outcome is otherwise rigged as a means to serve some end. For instance, an argument from order to the conclusion that the entire universe is the product of intelligence would suggest nothing about the use of means to satisfy ends. Instead, the universe may be the end in itself. If such an argument were sound there would be no need to suppose that the creator has employed contrivance, and so no limit on the creator's power is implied. But that still doesn't mean we have grounds to infer omnipotence, it merely shows that omnipotence can be compatible with the evidence. Hume seems to have been on to something. It is difficult to see how, from only finite evidence, one could ever be justified in concluding omnipotence. We'll consider an argument for the existence of an explicitly supernatural agent capable of violating natural laws in Chapter 15. But even this does not give us omnipotence. The ability to muster an infinite amount of some physical resource like energy or to violate a handful of natural laws is profound, but it is not omnipotence. We face similar hurdles with omniscience. With a finite sample of the designer's handiwork and uncertainty concerning the extent of the designer's powers, we have no grounds to infer such capacities as infinite foreknowledge. It looks as if we are left agreeing with Paley, who said that "a power which could create such a world as this is, must be, beyond all comparison, greater, than any which we experience in ourselves, than any which we observe in other visible agents."[15] The same goes for the creator's knowledge. But great is still not infinite.

I'll let Mill summarize the results of our deliberations:[16]

> These, then, are the net results of Natural Theology on the question of divine attributes. A Being of great but limited power, how or by what limited we cannot even conjecture; of great, and perhaps unlimited intelligence, but perhaps, also, more narrowly limited than his power: who desires, and pays some regard to, the happiness of his creatures, but who seems to have other motives of action which he cares more for, and who can hardly be supposed to have created the universe for that purpose alone.

---

[15] Paley 2006, 231.   [16] Mill 1998, 194.

10.5 Paley, Nieuwentyt, and why it matters

The final loose end I'd like to tie up concerns the relation between Bernard Nieuwentyt's now obscure *The Religious Philosopher* and William Paley's famous *Natural Theology*. You may have noticed some similarities between the arguments I discussed from each of these authors. If so, you are not alone. In 1833, by which time Paley's book had gone through many editions, an anonymous article appeared in an odd compendium called *The Book of Days* which contained essays, stories, accounts of curiosities, snippets of history and biography, and other short pieces to entertain readers with eclectic tastes.[17] The article was entitled "Bernard Nieuwentyt, the Real Author of Paley's 'Natural Theology'." It argues for precisely what the title suggests: that Paley plagiarized his famous book from his Dutch predecessor. The claim is not that Paley borrowed a common example, namely the example of the watch, but that the form of Paley's argument as well as the structure, details, and language of his presentation exactly parallel those of Nieuwentyt's earlier work. This initial accusation offers the following set of parallel quotations as evidence:[18]

> *Nieuwentyt.* So many different wheels, nicely adapted by their teeth to each other.
> *Paley.* A series of wheels, the teeth of which catch in and apply to each other.

> *Nieuwentyt.* Those wheels are made of brass in order to keep them from rust; the spring is steel, no other metal being so proper for that purpose.
> *Paley.* The wheels are made of brass, in order to keep them from rust; the spring of steel, no other metal being so elastic.

> *Nieuwentyt.* Over the hand there is placed a clear glass, in the place of which if there were any other than a transparent substance, he must be at the pains of opening it every time to look upon the hand.
> *Paley.* Over the face of the watch there is placed a glass, a material employed in no other part of the work, but in the room of which if there had been any other than a transparent substance, the hour could not have been seen without opening the case.

The plagiarism charge was elaborated by an author writing under the pseudonym "Verax" in the literary magazine *The Athenæum* in 1848.[19] In

---

[17] Anon. 1833.   [18] Anon. 1833, 197.   [19] "Verax" 1848.

1859, a detailed argument for plagiarism was made, this time under the author's real name.[20] It was written by the radical politician and historian of philosophy Robert Blakey.[21] Unless Blakey was plagiarizing the earlier accusations of plagiarism (one never knows), we can identify him as the author of all three works. (All three contain a variety of close commonalities, including the assertion that Paley's work is a "running commentary" on Nieuwentyt's.) I won't recount Blakey's case against Paley in detail; I leave it to the reader to compare the two books.[22] Rather, I will take it as a given that, in fact, Paley passed off as his own (or, more charitably, 'reworked') an argument that he took from Nieuwentyt. Why does this matter?

There are two reasons why it's worth pointing out the source of Paley's principal design argument. The first is simple fairness. In most modern literature, Paley's book is treated as though it were the beginning of natural theology. His argument, and in particular his treatment of the watch example, is held up as the epitome of classical design arguments. By 1830, virtually no one writing in English was citing Cicero, and even fewer were citing the innovative Nieuwentyt – everyone was citing Paley. So by some accident of history, Paley has received all of the credit properly due a small crowd of authors, Nieuwentyt chief among them. By pointing this out, I'm hoping to redistribute that credit a little. The second reason is that Nieuwentyt is much more explicit about the form of his argument than Paley is. Paley's work serves as a sort of mirror, reflecting whatever argument form – analogy, inference to the best explanation – each modern interpreter brings to it. While we want to make sure we consider each of these modern argument forms, we run the risk of missing an unconsidered possibility by imposing preconceived argument forms in this way. But by turning to Nieuwentyt, we can see unequivocally an argument distinct from either analogy or inference to the best explanation. In other words, by looking to Paley's source we discover an argument we would have missed if we had worked only with Paley's text.

---

[20] Blakey 1859.   [21] Hawkins 2012.

[22] For contemporary analyses that were sympathetic to Paley, see Anon. 1848 and Anon. 1911.

11 The modern likelihood argument

11.1 Probability, likelihood, and Bayes' Theorem

The first argument we'll consider from the modern era is a special sort of inference to the best explanation for which the vague term 'best' is given a rigorous formulation in terms of something called the 'Likelihood Principle'. To understand this version of the design argument, we need to be clear about some basic features of probability, so we'll begin this chapter with a quick review.

While it is easy to present the mathematical theory of probability – the probability calculus – it is difficult to give an account of just what probability is. This is because there are many different processes or features of the world that are well modeled by the probability calculus, just as many things can be modeled with algebra or geometry. There is likely no one thing that probability is *really* about, and in our discussion we'll see the calculus of probability applied in at least two different ways. On the one hand, probability describes processes that we think of as genuinely or inherently chancy – processes like the flip of a coin, the spin of a roulette wheel, or the decay of an atom. We tend to think that in at least some of these cases, there is no way in which the world must turn out – the coin could come up heads or tails – but the patterns in which such events are found to occur is well modeled by probability theory. The calculus lets us know what to expect. On the other hand, the probability calculus is good at modeling the ways we should change our beliefs about the world when we lack total information, at least if we want to be rational in certain ways. In this sense, probability is a measure of our degrees of belief. I won't try to argue for any one interpretation of the probability calculus (in part because I don't believe there is a unique interpretation), and an examination of the difficult questions surrounding the notions of randomness, chance, and 'degrees of belief' will lead us too far astray. Instead, I'll motivate the parts of the mathematical theory we'll need with an intuitive example involving a chance process. Once we have the math down, I won't stipulate an interpretation unless it is relevant.

An easy way to get a sense for the basic elements of the probability calculus is to consider the classic example of drawing marbles from an urn. Suppose we fill an urn with two kinds of marbles, glass (Gl) and plastic (Pl), each of which is also painted either black (Bl) or white (Wh). When the urn is full, it contains the following numbers of each sort of marble:

	Glass (Gl)	Plastic (Pl)
Black (Bl)	30	25
White (Wh)	20	25

If we now start drawing at random from the urn (say, by closing our eyes and reaching in, always taking care to stir the urn after returning a marble), how often do we expect to see a black glass marble? How often do we expect to draw a white marble? What about a plastic one?

These are the sorts of questions we can answer using the probability calculus. To do so, we begin by noting that a basic event consists of drawing a particular sort of marble from the urn. Looking at the table above, you can see there are four such events possible corresponding to the four types of marble. That is, you might draw a black glass marble or a white plastic marble, and so on. Let's label the four types of marble – the four basic events – as follows:

E_1:	black, glass
E_2:	black, plastic
E_3:	white, glass
E_4:	white, plastic

One would expect random draws to correspond to the proportion of each kind of marble, at least in the long run. In this case, we will identify the probability of each kind of basic event with these proportions. So, we have 30 marbles out of 100 total marbles, or 0.30 as the probability of drawing a black glass marble. We write this probability as '$P(E_1)$'. Likewise, we have the probability of drawing a black plastic marble is $P(E_2) = 0.25$. Similarly, $P(E_3) = 0.20$, and $P(E_4) = 0.25$. Notice that the probabilities of our basic events all sum to 1.

Often, we are interested in more complicated questions that involve collections of primitive events. For instance, we might ask for the probability that a marble drawn from the urn is white (a combination of E_3 and E_4) or the probability that a marble is plastic (a combination of E_2 and E_4). For such cases, the probability calculus tells us that the probability of the union of simple events is just the sum of their respective probabilities. So $P(Wh) = P(E_3 \cup E_4) = P(E_3) + P(E_4) = 0.45$ and $P(Pl) = P(E_2 \cup E_4) = P(E_2) + P(E_4) = 0.5$.

I won't attempt to systematically present the basic axioms and results of probability theory, but will instead use our simple example to introduce the relation we need, that of the 'conditional probability'. The idea here is to compute the probability of an event *conditional* on the occurrence of some other event. So for instance, we might ask for the probability that a marble drawn from the urn is white given that it is plastic. The conditional probability of some event A given the occurrence of some other event B is written "P(A | B)" and is defined to be:

$$P(A|B) = \frac{P(A \cap B)}{P(B)}$$

I didn't discuss it above, but the expression $P(A \cap B)$ just means 'the probability of A *and* B'. A simple way to think about the expression for conditional probability is just the number of events that are both A and B divided by the number of events that are B. It's like taking all the plastic marbles out of the urn and drawing only from this pile. Were we to do so, the probability of a marble being white is just the number of white and plastic marbles over the total number of plastic marbles. Thus in our example:

$$P(Wh|Pl) = \frac{P(Wh \cap Pl)}{P(Pl)} = \frac{0.25}{0.5} = \frac{1}{2}$$

As it turns out, we can use the definition of conditional probability to write one conditional probability in terms of another. It will become clear in a moment why we should want to do so. The result is:

$$P(A|B) = \frac{P(B|A)P(A)}{P(B)}$$

This is a result derived by Thomas Bayes in the eighteenth century,[1] and is the formula we will need in order to talk about the design argument. While not by any means the most important theorem Bayes proved in his paper (nor even an original result at the time), this equality is none-theless often referred to as Bayes' Theorem. Here's why we want to use it. Suppose I want to know the probability that some hypothesis, H, is true given some new evidence, E, I've uncovered. Using Bayes' Theorem we can write:

$$P(H|E) = \frac{P(E|H)P(H)}{P(E)}$$

In this expression, P(H | E) is the probability we're looking for; it's called the 'posterior probability' of the hypothesis, H. P(H) is whatever prob-ability we attributed to our hypothesis before we discovered the evi-dence E. It is referred to as the 'prior probability' of H. The term P(E) refers to the probability of having attained such evidence in the first place. The remaining term P(E | H) is known as the *likelihood* of H and represents the probability of finding evidence E given that the hypoth-esis H is true.

   If we treat the probability of a proposition as a measure of a rational degree of belief in that proposition, then we should believe the most prob-able hypothesis, and Bayes' Theorem tells us how to adjust our beliefs as we acquire new evidence. There is one problem with this general rule. In order to use Bayes' Theorem, we have to know the prior probabilities of both the hypothesis in question and the probability of attaining evi-dence. We generally don't have either of these things. If the hypothesis is, say, that of design, how can I have any rational degree of belief in the hypothesis absent any evidence? Even if I do, what on earth is the prob-ability of discovering a certain bit of evidence? For instance, what prob-ability should I assign to the discovery that the eye is well adapted for seeing? These problems push some to embrace a different approach, one that eschews nebulous priors and uses only the likelihood. Likelihood is

[1] See Todhunter 1865, ch. 14.

different. I may not know how probable design is, but I can say something about how probable it is that the eye is well adapted given that an intelligent designer produced it for the purpose of giving vertebrates excellent vision. But how can we use the likelihood alone?

11.2 The Likelihood Principle

The Likelihood Principle (or LP for short) began as a principle of statistical inference. According to an early defender, A. W. F. Edwards, "a particular set of data *supports* one statistical hypothesis better than another if the likelihood of the first hypothesis, on the data, exceeds the likelihood of the second hypothesis."[2] As with most of the mathematics of probability, this principle quickly escaped its original context, and has been promoted as a general rule for assessing evidence relative to any hypothesis, statistical, theological, or otherwise. Perhaps the most prominent proponent of this generalized version of LP is Elliott Sober. He provides the version of the principle we'll use:[3]

> **(LP)** Observation O supports hypothesis H_1 more than it supports hypothesis H_2 if and only if $P(O|H_1) > P(O|H_2)$.

Notice that LP does not say which hypothesis is true or probably true. Nor does it say which hypothesis we should believe or how we should respond to the evidence. All that LP can tell us is how hypotheses rank with respect to one another in light of a given bit of evidence.[4] That is, LP tells us in which direction we ought to adjust our beliefs concerning H_1 and H_2, but not which one should win out in the end. It doesn't give us enough information to decide which, if either, is probably true. That should be unsurprising. After all, we left out information essential to computing a posterior probability. The fact that LP is merely comparative in this way is both a strength and a weakness. It's a strength because LP lets you say something about your hypotheses given the evidence in those cases in which we have no information about prior probabilities. LP says that the hypothesis with the greatest likelihood is most favored by the evidence, and we should

---

[2] Edwards 1972, 30.   [3] Sober 2003, 29.

[4] In later work, Sober (2008) has defended a 'Law of Likelihood' which adds to LP a quantitative measure of the support for one hypothesis over another provided by a piece of evidence.

increase our belief in that hypothesis at the expense of the others. But the strictly comparative nature of LP is also a weakness. We generally don't care in what direction the evidence points – we want to know what's true! As we'll see below, LP needs to be supplemented in order to function in an argument for any particular hypothesis.

11.3 An example: probability vs. likelihood

Likelihood is in many ways a counterintuitive concept. So before we attempt to apply LP to the design argument, let's take a moment and work through an example that highlights the distinction between likelihood and posterior probability. Suppose that a patient walks into the office of Dr. Harvey exhibiting symptom S. Out of the 30,000 patients Dr. Harvey has seen in his career, this is the first patient to have S. Now, Dr. Harvey knows of two diseases that have been reported to produce S. Let's call them disease X and disease Y. In every known case of disease X, the patient developed symptom S. But X is a rare disease, afflicting only 1 in 3 million people. Disease Y is much more common, afflicting 1 in 1,000 people. But in cases of Y, only 1 in 100 results in symptom S. Which explanation should Dr. Harvey favor?

Let's first consider which hypothesis is favored by the evidence at hand according to LP. To answer this question, LP says we need to compute the *likelihood* of disease X and the *likelihood* of disease Y in light of the symptom S. That is, we have to find $P(S \mid X)$ and $P(S \mid Y)$. In this case, we can estimate the likelihoods if we assume that the frequencies I gave in describing the problem are good estimates of the true probabilities. So, for instance, $P(S \mid X)$ is just the number of patients who have disease X and develop symptom S, divided by the total number of patients that have X. Since everyone who gets X has S, we just have $P(S \mid X) = 1$. Likewise, $P(S \mid Y) = 1/100 = 0.01$. What do these numbers mean? Well, LP tells us that the evidence of symptom S favors the hypothesis that the patient has X over the hypothesis that the patient has Y.

But what about the posterior probabilities? In this case, we do have enough information to assign meaningful, objective prior probabilities. So we can find the *probability* that the patient has disease X given that he has symptom S, $P(X \mid S)$, and the *probability* that the patient has disease Y given that he has symptom S, $P(Y \mid S)$. Notice that the disease and the symptom have switched sides in the expressions for probability. To find numerical values for these probabilities, we just use Bayes' Theorem:

$$P(X|S) = \frac{P(S|X)P(X)}{P(S)} = \frac{(1)\left(\frac{1}{3} \times 10^{-6}\right)}{\left(\frac{1}{3} \times 10^{-4}\right)} = 10^{-2} = 0.01$$

$$P(Y|S) = \frac{P(S|Y)P(Y)}{P(S)} = \frac{\left(\frac{1}{100}\right)\left(\frac{1}{1000}\right)}{\left(\frac{1}{3} \times 10^{-4}\right)} = 3 \times 10^{-1} = 0.3$$

Notice that there is a big difference in the posterior probabilities and a difference as to which hypothesis comes out on top. Even though, according to LP, the evidence favors X over Y, the posterior probabilities tell us that we should still believe the patient has Y. Why is that the case? How could the likelihoods be so misleading? The problem is not with LP per se, but rather the intuitive appeal of mistaking LP for more than it is. The evidence does favor X over Y in the sense that, if the patient has X, it is no surprise at all that he has S. If the patient has Y, then symptom S is just an unlikely coincidence. As far as our evidence goes, X explains everything with no coincidences! But that's just what one piece of evidence tells us – LP is a recipe for ranking hypotheses with respect to evidence. We make a huge mistake if we confuse this ranking with a way of directly choosing what to believe. In this case, the disease X is so rare that it is much more probable the patient has a somewhat unusual symptom of a much more common illness. This mistake of ignoring known prior probabilities has a name – it's called the Base-Rate Fallacy and it's something doctors have to watch out for. In fact, someone very dear to me was almost lost because a team of medical experts committed the Base-Rate Fallacy. If our fictional Dr. Harvey is to make use of all the information available to him and avoid fallacies, he should skip LP, use Bayes' Theorem, and suspect disease Y. The more general lesson is that there is a big difference between likelihood and probability. While a hypothesis may have a very great likelihood in light of a new piece of evidence, its prior improbability may be so great as to make the posterior probability quite low.

If it can be so misleading, why would we ever use LP? The answer sup-
porters of LP give is that we use likelihoods when that's all we've got.
Suppose, for instance, that we knew nothing about the frequency with
which X and Y occur in the population. According to 'subjectivists', probability is a measure of one's degree of belief that need not be tied to an
empirical frequency. Many adopt the rule that if we are equally ignorant
concerning two hypotheses, we should grant them equal probability. This
is the so-called 'Principle of Indifference'.[5] If our prior probabilities are
equal for a pair of propositions, then it turns out that the ratio of likelihoods is equal to the ratio of posterior probabilities. In other words, if we
assume that $P(X) = P(Y)$, then whichever one has the highest likelihood
given the evidence also has the highest posterior probability. Of course,
you might reject the Principle of Indifference.[6] In that case, a defender of
LP might still argue that even if we lack any evidence regarding the prior
probabilities of hypotheses, then the likelihood ratio is still informative –
it tells us which way the odds in favor of one hypothesis are shifted from
what they were before we learned the evidence. If the ratio of likelihoods
favors H_1, then the posterior odds in favor of H_1 are greater than the prior
odds. Of course, since we don't know what the prior odds were, we cannot
tell which hypothesis has the greatest posterior probability. But LP does
make the most of the information available to us by indicating which way

[5] See, e.g., Edwards 1972, 55.
[6] There are a variety of reasons to find the Principle of Indifference suspect. Most criticisms involve the question of what precisely we are supposed to be indifferent about.
A brief example due to von Mises (1981, 77) makes the point nicely. Suppose we have a
pitcher filled with a mixture of wine and water. All we know in advance is that there
is at least as much water as wine and at most twice as much water as wine. What's
the probability that the ratio of water to wine is at most 3/2? Well, we know the ratio
lies between 1 and 2, and so the Principle of Indifference tells us that the probability
must be 1/2 since exactly half of the possible ratios are greater than or equal to 3/2.
But here's the problem. I could just as well have given the available information in
terms of the ratio of wine to water. In that case, the possible ratios range from 1/2 to
1. Asking for the probability that the ratio of water to wine is at most 1.5 is the same
as asking for the probability that the ratio of wine to water is at least 2/3. This time,
the Principle of Indifference tells me that the probability is $(1–2/3)/(1/2) = 1/6$! Since
there is no a priori reason to prefer one description over the other, the Principle of
Indifference cannot give a unique answer; the principle is incoherent. For a recent
discussion of this problem, see Shackel 2007. For a recent defense of the Principle of
Indifference, see Norton 2008.

the posterior odds are shifted by the evidence. This is true whether or not one is a subjectivist about probability. The best we can do in this case, says the supporter of LP, is to determine which of the available hypotheses the total evidence moves us toward, and reserve judgment as to which is most likely true.

11.4 A likelihood argument for design

What can it mean to give a likelihood argument for design? After all, a design argument is supposed to support a particular hypothesis, namely that the universe has a designer. But I keep insisting that LP is merely comparative – it tells you which of two hypotheses a particular body of evidence favors, not which is true. To draw a conclusion of design, we'll need a little more than LP itself. The strategy is to add enough to get a plausible inference to the best explanation (IBE). You may recall from Chapter 8 that in an inference to the best explanation, we infer from a set of possible explanations for the available evidence that which is the 'best'. Now, LP gives us a way to say exactly what we mean by the best: it's whatever hypothesis has the highest likelihood. Of course, if we use likelihood as our measure of 'best', then we risk running into the Base-Rate Fallacy. So we must further assume that all hypotheses start off on an equal footing with roughly equal prior probabilities. In other words, we must adopt the Principle of Indifference.

Now we're ready to use likelihoods to give a design argument.[7] The evidence in question is the remarkable adaptedness of living things for surviving in the environments in which they find themselves. In other words, we're going to focus on exactly the same kinds of facts that enthralled Ray, Derham, Paley, Darwin, and probably everyone else who has considered natural history at any length. These facts will be used to assess two hypotheses: first, that the exquisite adaptations of organisms are the product of chance processes, and second, that organisms are the result of intelligent design. Let's label each of these propositions as follows:

---

[7] See Sober 2000 and 2003 for a likelihood version of the design argument. The argument I present diverges somewhat from that of Sober, who does not consider the need for the Principle of Indifference.

E_0: Organisms are exquisitely adapted to tasks that keep them alive and perpetuate the species.

H_1: The traits of organisms are the result of intelligent design.

H_2: The traits of organisms are the outcome of chance processes.

Now, given that H_1 is true, it is very probable that organisms would exhibit adaptations. That's pretty much what it means to be designed. But if H_2 is true, it would seem to be very improbable that organisms would have such finely adapted structures. While we cannot put precise numbers on these likelihoods, we can say that $P(E_0 \mid H_1)$ is much greater than $P(E_0 \mid H_2)$, and so the evidence from biology favors design over chance processes. Given the rest of our assumptions, we should believe that organisms are the product of design.

11.5 Some objections considered

How strong is this version of the argument? One way we might start feeling out the likelihood argument is to consider how it fares against the more powerful objections leveled at the argument from order. Thinking back to Hume's *Dialogues Concerning Natural Religion*, the strongest objection rested on the claim that we cannot empirically justify the assertion that a particular property is always and only the result of intelligent agency because there will always be cases – such as living things – for which we are unable to witness the creation of the property. LP dodges this concern tidily. To invoke LP, we need not know that adaptedness can only be produced by design. We need only know that if the thing was designed then it would exhibit adaptedness with high probability. As long as we know the likelihood associated with the design hypothesis, we don't need to have witnessed any such events of creation. LP lets you hypothesize whatever you like. As long as the hypothesis entails at least an approximate probability for the evidence, that's enough to favor it in a likelihood argument.

But the argument from order is vulnerable in other ways. Recall that the argument typically involves rejecting randomness in favor of intelligence as the source of some striking property like order or complexity or goal-directedness. Hume suggested, as did the Epicureans before him, that given sufficient time, chance really could produce the sort of order we see. More accurately, the chance encounter of atoms initially in a chaotic state could result in an ordered world as long as atoms interact with one another according to natural laws. As Hume would have it, once the universe of swirling atoms stumbled across a configuration similar to that in which we find it, the unintelligent laws of nature would guarantee that it stays in what we would call a 'dynamically stable' state – atoms move all the time, but stuff stays roughly the same. If this were true, then we would not be able to infer intelligent agency from the presence of order – we would first have to rule out a process like that which Hume describes. Could we make a similar objection in the case of the likelihood argument? Let's add a third hypothesis:

> H_3: The traits of organisms (and the universe as a whole) are the product of a process involving chance, the laws with which atoms blindly interact with one another, and a great deal of time – after a very long time, the universe eventually stumbled across a configuration that is dynamically stable.

Now we must ask whether the likelihood of this hypothesis changes our conclusion. To answer this, we need to know how $P(E \mid H_3)$ compares to the likelihoods of the other two hypotheses. According to Sober, it would be a mistake to think that $P(E \mid H_3)$ is any bigger than $P(E \mid H_2)$. H_3 sounds plausible only because of the presumably unlimited time available for the universe to find a dynamically stable configuration. But, says Sober, to argue for a large likelihood for H_3 is to commit the 'Inverse Gambler's Fallacy'.[8] To see this consider a less complicated example. Suppose that a gambler walks into a casino and immediately witnesses the roll of double-sixes. This will be the fact we wish to explain:

> E_1: A double six was rolled.

Two hypotheses to explain this roll come to mind:

> H_4: This is the first toss of the dice tonight.
> H_5: Many throws of the dice have been made to this point.

The gambler might think that, if many rolls have been made – if people have been gambling for a while – then he is rather likely to witness a double-six. Thus, H_5 has a high likelihood. On the other hand, the likelihood

[8] Sober 2003, 37; also Hacking 1987.

of H_4 is just 1/36, the probability of rolling a double-six in one roll. This is analogous to Hume's Epicurean argument in the sense that Hume thinks the likelihood of the universe attaining its current configuration is high if lots of configurations have been tried already. But according to Sober, the gambler and Hume have made analogous mistakes.

The fallacy for the gambler lies in the fact that E_1 is ambiguous as I have stated it. If we understand E_1 as the claim that a double-six was rolled sometime during the evening, then the gambler is correct. The likelihood that the roll occurs late in the evening is high, much higher than that it occurs on the very first roll. But the information we have available to us is much more specific. We have evidence of rolling a double-six on one *particular* roll. Since tossing the dice is random in the way a fair coin is random, the probability of an outcome on one roll is independent of the outcomes on previous rolls. Processes with this feature are called 'Markov processes'. On this more precise reading of our evidence, $P(E_1 \mid H_4) = P(E_1 \mid H_5) = 1/36$. In that case, we cannot favor one hypothesis over the other. The implication for Hume's hypothesis is supposed to be that if we account for the evidence we actually have – that we live in this universe at this time and not some other – then the outcome is still very improbable. That is, if the universe assumes states at random in such a way that it is a Markov process, then the likelihood is exceedingly low of the universe stumbling into the configuration we see it to have at this very moment, no matter how much time the universe has already existed.

The analogy, however, is inapt. The gambler commits the fallacy when he confuses a hypothesis that makes a general proposition very likely with one that makes a specific proposition probable. No such confusion seems to be involved with Hume's objection. When we are speaking of the adaptations of organisms, both design proponents and their critics are talking about the general occurrence of adaptation, not the specific occurrence of the particular adaptations we happen to see and the particular times at which they happen to be seen. If we insist on explaining the narrow fact, then neither design nor chance has a high likelihood. Surely a designer would make the parts of organisms well adapted for functions necessary to the survival of the organism. But what are the odds that God would choose to make the vertebrate eye precisely as it is and at the time at which it first appeared in this world. At the very least, it's not obviously a probable outcome. Furthermore, Sober's characterization of the chance hypothesis isn't quite fair to Hume. He suggested that the universe only had to stumble across a configuration like ours one dynamically stable piece at a time. That is, once one self-sustaining process gets started (like the formation of a swirling cloud of gas held together by gravity), it sticks around while the rest of the atoms continue randomly jostling about. In other words, successive states are not independent of one another like the rolls of the dice. It is, true, however, that Hume doesn't really give us any specifics. Why would the laws tend to capture dynamically stable states out of chaos? What are the laws exactly? What is the likelihood they would result in organisms at all, whether or not they are adapted? Thus, whether or not we accept Sober's analysis, Hume's Epicurean rebuttal doesn't seem to get much purchase on the likelihood argument.

But Darwin's does. In *The Origin of Species* he presents a wide variety of phenomena which are to be expected if evolution by natural selection is true, but that would be quite surprising if species were created de novo by an intelligent agent. For instance, he points to patterns in biogeography, the fossil record, and taxonomy. Though Darwin doesn't put it in these terms, a large portion of his arguments can be cast as likelihood arguments. The idea is to introduce as hypothesis H_6 the claim that species arise by evolution through natural selection. Darwin argues that $P(E_0 \mid H_6) > P(E_0 \mid H_1)$ and so, according to LP, we should favor H_6 on the basis of E_0 and the rest of the available facts about biology. Of course, a lot hinges on just how probable Darwin's theory (or its modern counterpart) makes the evidence cited and on whether we might conceive of a more competitive design hypothesis. But with what's currently available, natural selection dominates. Insofar as likelihood arguments are convincing, it looks as if we should favor evolution.

11.6 Sober's sticking point: the objection from abilities

Sober thinks we can defeat the design argument without appealing to the Darwinian hypothesis. His criticism has to do with the way likelihoods are determined and what counts as a hypothesis. More particularly, it has to do with the conditions necessary for an objective likelihood to be determined for a given hypothesis. In order for there to be a fact of the matter whether the chance hypothesis H_2 has a high or low likelihood, we have to stipulate what is meant by a chance process with enough detail to infer how often such processes result in the observation under consideration. One way to do so is by insisting that chance processes are those that obey the Markov condition, and that all possible arrangements of atoms in space are equally likely at any given instant of time. This makes the probability of any one configuration vanishingly small. Of course, this is a very implausible scenario, and much more radical than that which the Epicureans defended. But the point is that the added detail is necessary to get a concrete likelihood. In the more realistic case of medical diagnosis we face a similar problem. To put an objective number on the likelihood of a given disease, we would have to be able to say just how often a given symptom manifests in patients with that disease and then assume that all patients have an equal probability of displaying the symptom given the disease. That is, we need to assume some things about the nature of the process in question that allow us to apply the probability calculus in a definite way. As Sober puts it, every hypothesis must be supplemented with auxiliary assumptions that give us enough information to compute a likelihood.

In the case of the design hypothesis, a likelihood can be computed (or at least estimated) only if we can say something about the abilities and goals of the designer. Otherwise, we cannot say anything about how many outcomes are possible under this assumption, and what proportion of these match the evidence at hand. As Sober puts it, taking the extraordinary adaptedness of the vertebrate eye into consideration: "[E]ven if we assume that the eye was built by an intelligent designer, we can't tell from this what the probability is that the eye would have the features we observe."[9] Without knowing the designer's goals and abilities with respect to enacting these goals, we cannot assess how probable any particular result is. This, argues Sober, is an insurmountable difficulty for the likelihood argument.[10]

Of course, you might wonder why we can't just supplement the design hypothesis with suitable auxiliary assumptions. For instance, we might consider the augmented hypothesis that God wants organisms to thrive and perpetuate. Or we could add something else about God's goals and abilities to the hypothesis, or entertain many hypotheses, one for each

---

[9] Sober 2003, 38.

[10] Sober gives a more careful, technical treatment of this objection in terms of the probability calculus. However, the details add little to the force of the objection and so I have not recounted them here.

plausible combination of goals and abilities. But Sober thinks this cannot be done without somehow begging the question of design: "One needs *independent* evidence as to what the designer's plans and abilities would be if he existed; one can't obtain this evidence by *assuming* that the design hypothesis is true."[11] Since we don't have any such independent evidence (what would it look like?), we can't provide the necessary information to carry through a likelihood argument.

But this objection seems less than even-handed. The same problem with auxiliary assumptions applies to the so-called chance hypothesis and to the Darwinian hypothesis favored by Sober. Let's just consider 'chance'. As the details of Hume's version of the Epicurean model demonstrate, there is more than one way to add to the vague hypothesis of 'chance'. We might, for instance, suppose that each state of the universe is completely independent of that which came before, as Sober does. In another, like that assumed by Bentley as the chance hypothesis, atoms which begin in an initially random configuration proceed to interact with one another in a lawlike fashion. The point is that there are many ways to supplement chance hypotheses, too. Which are we to choose? We have no independent evidence of which mechanism we ought to be considering.

Sober seems to be invoking extra rules to suit his aim. To begin with, there is no clear logical distinction between hypotheses and auxiliary assumptions, but a plausible position for any supporter of LP is that a hypothesis is any proposition which yields a definite probability in light of the evidence. In that case, Sober's objection is not really about picking 'auxiliary assumptions' but rather identifying allowable hypotheses. But LP tells us nothing about what counts as an acceptable hypothesis. Nor does the Principle of Indifference. So it seems we have to either entertain them all or risk begging the question in favor of one or another conclusion. What this objection really shows is not so much a problem with the likelihood design argument as with likelihood arguments in general. The problem is that we can always cook up a hypothesis which makes the evidence certain and thus has the highest possible likelihood. So, for instance, we might hypothesize that there exists a powerful God who wanted animals to have vision just like ours and who prefers to make eyeballs with a structure just like ours. This is obviously a silly and ad hoc hypothesis, but it

[11] Sober 2003, 38.

does give the evidence the highest possible likelihood. So what is there to bar us from introducing it? And doesn't LP compel us to embrace this hypothesis? Sober has no general principle to offer here. Without one, the worry doesn't discriminate between the design hypothesis and others.

11.7 Are likelihood arguments ever convincing?

Given the problems I've raised for likelihood arguments, it's reasonable to ask whether LP can ever provide the basis for a strong argument. Sober provides an example of what he takes to be a successful application of LP. The first is drawn from the personal recollections of the biologist John Maynard Smith. In the Second World War, one of Maynard Smith's jobs was to inspect German warehouses in Ally-seized territory. The warehouses contained all sorts of machines, parts, and materiel for the Nazi war effort. Many of the machines Maynard Smith and his colleagues encountered in this context served purposes that were entirely unknown to them. Nonetheless, they were able to recognize that these machines did in fact serve *some* purpose – they were the products of intelligent design.

Now, Sober interprets this uncontroversial inference as an application of the likelihood argument. Given the known (very large) set of goals and abilities German engineers are known to possess, a rather firm likelihood can be attached to the hypothesis that the machines were designed to have the features they were found to have, and that this likelihood far exceeds the likelihood that the machines were accidental conglomerations of scrap, random assemblages of parts, or naturally occurring formations. Thus, says Sober, Maynard Smith and his colleagues were able to infer design by making a likelihood argument.

But consider which aspects of the machines gave away their designers: "[T]he items in Maynard Smith's warehouse were symmetrical and smooth metal containers that had what appeared to be switches, dials, and gauges on them."[12] All of these are properties of objects which, in our experience, are *always and only* associated with intelligent human agency – they comprise what we might call 'marks of design'. That is, we could argue for design in this case in the following way:

---

[12] Sober 2003, 39.

P1: The object exhibits a great deal of order: it is symmetric, of simple geometric form, is composed of an alloy not found in natural rock, it has what appear to be switches and gauges, etc.

P2: Order of this kind always results from human agency.

C: The object is the result of human agency.

Here, the premise that such marks of design are always indicative of human agency is strongly (inductively) supported by experience. We need not appeal to any sub-argument or overarching theories of human intelligence. Nor do we need to attempt to assess the probability that an object would have any or all of these exact properties given it was designed by humans or formed naturally, etc. Given the uncertainty involved in assessing those probabilities, this simple argument from order seems to offer a stronger argument than any we could construct from likelihoods. I suggest, then, that, contrary to Sober, we should view Maynard Smith's wartime inference as a simple argument from order where the link between properties like symmetry and human intelligence is inductively justified in a transparent way. No LP, no Principle of Indifference.

A similar analysis pertains for two other modern examples of successful design arguments. The first concerns the Antikythera Mechanism, which I described at the start of this book. It's the object that was discovered in 1901 within the remains of a shipwreck off the coast of the Greek island of Antikythera. It dates to the end of the second century BCE. At first, it was anything but apparent what its purpose might have been, but it was immediately recognized as the product of intelligent human agency. Recent studies using advanced imaging to read inscriptions on the remaining fragments of its case have confirmed that the device was used to compute with extraordinary precision the occurrence of solar and lunar eclipses, as well as the positions of the visible planets.[13]

But on what grounds was the original inference drawn? How could modern archeologists be so sure that an oxidized lump of copper and rotten wood is an artifact at all? The answer, I think, is that the device bears clear marks of human design. Just like the mysterious German machines that Maynard Smith encountered in the Second World War, the Antikythera Mechanism is made predominantly of a metallic alloy that does not occur

[13] Freeth *et al.* 2006.

naturally. Furthermore, it contains geometrically simple, symmetric parts in the form of wheels and gears, some of which have prime numbers of teeth! Even were there no writing on the thing at all, the various properties just mentioned would be sufficient to support a strong inference to design.

My second example comes not from under sea, but from outer space. In late 1967 while monitoring the data from a newly constructed radio telescope, S. Jocelyn Bell Burnell noticed a remarkable signal. "As the chart flowed under the pen I could see that the signal was a series of pulses, and my suspicion that they were equally spaced was confirmed as soon as I got the chart off the recorder. They were 1⅓ seconds apart."[14] Nothing like this had ever been observed in radio or optical frequencies in astronomy to that point. Naturally, the first thing the team considered was a stray, man-made signal. However, they soon eliminated all plausible man-made sources, and had to confront an extraordinary possibility: "So were these pulsations man-made, but made by man from another civilization?"[15] That is, the next hypothesis the team considered was that the precise, regular signal they were witnessing was the product of a non-human intelligent agency, presumably an extraterrestrial technical civilization.

Why was this a plausible inference? In his discussion of the SETI (Search for Extraterrestrial Intelligence) project, Sober again casts the inference to alien intelligence in terms of likelihood. However, I would argue that the strongest support for the possibility of intelligent agency derived from the simple fact that the only regular signals we know of – the only radio sources displaying this sort of order – were human-produced. Once humans were eliminated from the set of possibilities, a similar argument from order could be advanced for very human-like intelligent agents on other worlds. Of course, in this case the design hypothesis turned out to be false and the argument unsound. This just shows that confidence in an argument from order extends only as far as our certainty that all sources of order have been accounted for. That's why Darwin's work was so destructive to Paley's argument – Darwin discovered an alternative source of order that would have to be eliminated before design could be concluded. How do we know that the design hypothesis is false in the case of the mystery pulsations from space? This possibility was eliminated because if the signal

[14] Burnell 2004.   [15] Burnell 2004.

was coming from intelligent life forms on another planet, "then the pulses should show Doppler shifts as the little green men on their planet orbited their sun. Tony Hewish started accurate measurements of the pulse period to investigate this; all they showed was that the earth was in orbital motion about the sun."[16]

What lesson should we draw from these examples? Even if it's possible to justify LP and the Principle of Indifference it doesn't seem that we gain anything. In those instances in which it's possible to extract objective probabilities for evidence given a particular design hypothesis, it is also possible to offer a simpler argument from order. In short, the argument from order is a stronger approach.

[16] Burnell 2004.

12 Intelligent design I: irreducible complexity

12.1 Introduction to the modern arguments from biology

Modern design arguments that draw upon biological facts are typically grouped under the heading of 'Intelligent Design' (or ID for short). The name is apt, if unimaginative. What ID arguments have in common is the relatively modest goal of establishing the existence of a non-human designer without attempting to establish any characteristics of this designer. Additionally, ID arguments purport to abide by the norms of empirical science. Within ID, there are two dominant strands of argument, each appealing to a different feature of organisms to infer design. First, there are those arguments which appeal to 'irreducible complexity', a term coined by Michael Behe.[1] The notion of irreducible complexity is tightly bound up with the idea of gradual evolution – the central idea is that irreducibly complex systems cannot be produced by a process of gradual evolutionary change. Arguments based on irreducible complexity generally involve an attempt to identify biological systems with this property, and then to argue from the existence of these systems to the likely intervention of an intelligent agent. In this chapter, we will consider just what the property of irreducible complexity is supposed to be and what sorts of design argument it might support. In the next chapter, we'll turn to an examination of the second major strand of biological design argument: that which appeals to 'specified complexity'. As we'll see, both sorts of argument share a similar structure – a structure with very deep roots in the history of design arguments. This common structure is buried under two different sets of jargon. We will begin digging it out in this chapter by scrutinizing the notion of irreducible complexity.

[1] Behe 1996.

12.2 Irreducible complexity and the argument from order

Irreducible complexity, if its advocates are right, is a very special property – it is a marker or footprint of intelligent agency. But what exactly
is this property? What does it mean to say that a physical system (e.g. a
mousetrap, a laptop, or an elephant) is irreducibly complex? In his original
work on the subject, Behe[2] – who remains the principal defender of the
argument from irreducible complexity – offers the following definition:[3]

> Irreducibly complex system:
>
> [A] single system which is composed of several well-matched, interacting
> parts that contribute to the basic function, and where the removal of any
> one of the parts causes the system to effectively cease functioning.

To unpack this definition, let's consider an instance of irreducible complexity. Behe likes to use the example of a mousetrap, but for those of you
unfamiliar with the art and practice of killing rodents, perhaps a stapler would provide a more familiar illustration. The typical office stapler
would, if anything does, conform to Behe's notion of an 'irreducibly complex' system. First, there are lots of well-matched interacting parts that
contribute to the basic function of driving staples through stacks of paper.
I say the parts are well matched in the sense that a metal flange or "tooth"
mounted on the upper part of the stapler is placed and oriented just right
so as to pass through a narrow slot in the chamber that holds the staples
whenever someone presses down on the device. That slot and the tooth
are made of sturdy enough materials to sustain the stress necessary to
push a staple through multiple sheets of paper. Depressions on the end of
the baseplate beneath the tooth and slot are curved and placed just right
to guide the ends of the staple inward after they have passed through the
paper. A spring pushes staples forward in their chamber so that one is
always beneath the tooth. If all this sounds like the way in which the parts
of an organism are adjusted to one another that is precisely the point.

---

[2] Despite the prominence of the argument from irreducible complexity in the literature on ID, Behe is – to the best of my knowledge – the only author to publish a developed defense of the concept and the associated design argument. For this reason, I
focus solely on Behe's work.
[3] Behe 1996, 39.

Returning to the stapler, it is the case that if we remove any of the parts mentioned above the stapler would cease to function. For instance, if you took out the spring, no staples would be pushed into position below the tooth, and so none would be driven into paper. They would sit inert at the back of the chamber. Likewise, if you left the spring but removed the tooth, staples would never be ejected. If you don't believe me, it's an easy experiment to try at home. Because each of the parts is necessary for and finely adjusted to accomplish the function of stapling paper, a stapler is an instance of what Behe means by 'irreducible complexity'.

To this point, we have left a crucial component of Behe's definition unexamined: the notion of 'basic function'. Objects do not come with labels indicating their basic function. How do we determine, out of all the things an object actually does and all the ways an object potentially inter-acts with the world, which is its basic function? In the case of the stapler (and Behe's mousetrap) we know in advance for what human *purpose* the thing was designed, and so it is intuitively obvious what its basic function is. But think of the Antikythera Mechanism described in Chapter 1. The first archeologist to notice the device in the scrapbin had no idea what the machine was supposed to do. How, then, could we determine its 'basic function' to decide whether it is irreducibly complex? Prior to being fished off the ocean floor, the Mechanism served as an excellent substrate for marine invertebrates. Is that its primary function? If you get a good grip on the largest part of the Mechanism, you could use it to crack walnuts. Is that its primary function? These suggestions sound silly. We know from textual and other evidences that the Mechanism was intended to compute astronomical data. But physical systems that were not crafted by human hands do not come with inscriptions telling us what they are for – we need a principled way of deciding which of the things the Mechanism does (or did before being sunk) are its basic function(s) and which are just things it happens to do. In Behe's example of the mousetrap and my example of the stapler, basic function happens to line up with human purposes. But we don't want to define 'basic function' in terms of the 'principal purpose of the designer' because then ascribing irreducible complexity to a system would be begging the question – it would assume that a designer with intent exists and has crafted the system.

One way to try and resolve the problem is to 'naturalize' the notion of function – to give a definition of 'function' that captures what we seem to mean when we speak of, say, the 'function of the heart' but which does not appeal to the intentions of any intelligent agent. There is a very large literature on naturalizing functions in the philosophy of biology. In this literature, one can find two basic approaches to the problem. First, one can appeal to the evolutionary history of a particular system like the eye. In this approach, one identifies the function of a system with those causal properties for which it was favored by natural selection.[4] Under this account, the function of an eye is to see because our ancestors were favored by natural selection in part for their ability to see. However, if irreducible complexity is to be a marker for intelligent agency, it cannot appeal to natural selection – a decidedly unintelligent cause. If we spell out function in terms of natural selection, then every irreducibly complex system would by definition be a product of evolution, not intelligent agency.

We might instead spell out the idea of a function according to the second major strategy in the literature on naturalizing the concept: we could try to spell out the function of a system in terms of the capacities of that system which contribute to the capacities of a larger system – call it the 'super-system' – of which it is a part. According to this approach, we might say that the function of a system x is any capacity which x has to produce effects that in turn explain why the super-system which contains x has some other capacity y.[5] This sounds abstract, but it is really straightforward. It is just the claim that the function of a system is whatever it does that makes the larger super-system containing it do something else. So for instance, we could say that a function of the heart is to pump blood because the capacity of the heart to pump blood contributes to the capacity of the circulatory system to distribute nutrients and oxygen to the body. Or, to refer back to the example of the stapler, one might say that a function of the base plate is to bend the staples since this ability contributes to the capacity of the stapler to bind papers together.

We can use this approach to naturalize the notion of function as it appears in Behe's definition of irreducible complexity by thinking of the 'basic function' of the system under scrutiny as whatever capacity of that system – call it x – contributes to a capacity of the super-system in which

---

[4] This is roughly Larry Wright's (1973) notion of function as applied to evolutionary biology. See Godfrey-Smith 1993.

[5] What I've given here is a version of the definition of 'function' presented by Robert Cummins (1975). See also Godfrey-Smith 1993.

the system is found. Saying that such a system is irreducibly complex is just saying that to do x, each of the parts making up the system are necessary. After all the dust settles, here is what the revised definition looks like when these pieces are put together:

Irreducibly complex system (2):

A system with a capacity to do x which is composed of several interacting parts that each have capacities which contribute to the system's ability to do x, and where the removal of any one of the parts causes the system to lose its capacity to do x.

This new definition eliminates the undefined term 'function' in favor of the objective, causal capacities of the system and its parts. However, it comes at a cost: irreducible complexity is now explicitly relative to the capacity x. We defined 'basic function' in such a way as not to beg the question of design, but we have no principled way of deciding which capacity of the super-system – or even which super-system – is the relevant one, and thus no way of deciding what x is for the system under consideration. Any system has lots of capacities, and the same system can be irreducibly complex with respect to some of these capacities and yet *not* irreducibly complex with respect to others. For instance, office staplers are – according to our revised definition – irreducibly complex with respect to the capacity of staplers to fasten papers together. Each part of the stapler has a set of capacities that contribute to this overall capacity, and if you take away any of its parts the stapler loses the capacity to fasten papers together. However, with respect to its capacity to bind posters to a corkboard, the stapler is not irreducibly complex – we could remove the entire bottom half of the stapler without affecting its capacity to staple posters to the wall. As we'll see, this relativity of irreducible complexity is a problem for a key claim in Behe's argument for a designer.

Behe's argument for design begins by establishing that some biological systems are irreducibly complex. One of Behe's favorite examples of such a system is the bacterial flagellum. A flagellum is a long appendage which protrudes from the exterior of many sorts of bacterial cells and is used for locomotion. Figure 12.1 depicts a bacterium with a number of prominent flagella. When spun along its long axis, the flagellum is twisted into a rotating corkscrew that pushes the bacterium through the fluid in which it lives. In order to rotate the flexible shaft of the flagellum, the bacterium

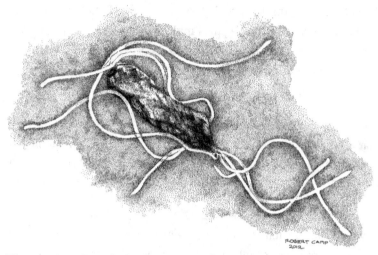

Figure 12.1 Drawing of an electron micrograph of *H. pylori* possessing multiple flagella (negative staining). Illustration by Robert Camp, based on an image created by Professor Yutaka Tsutsumi, M.D., Department of Pathology, Fujita Health University School of Medicine

is equipped with a complex molecular motor – a relatively large assembly of proteins that operates like a rotary motor, using a difference in acid concentration across the cell's membrane to drive the rotation of the shaft. For details, I refer the reader to Behe's own apt description of this system.[6] It suffices here to note that if any one of a handful of protein parts is removed from the flagellar motor, the shaft cannot be rotated. Behe claims that rotation is clearly the 'basic function' of the bacterial flagellum (i.e. the capacity we should be worried about) and argues that, since removal of any one part of the molecular motor deprives the flagellum of the capacity to rotate, the whole system is irreducibly complex.

If we concede that the flagellum is an irreducibly complex system, what follows? Here is how Behe puts it:[7]

An irreducibly complex system cannot be produced directly (that is, by continuously improving the initial function, which continues to work by the same mechanism) by slight, successive modifications of a precursor system, because any precursor to an irreducibly complex system that is missing a part is by definition nonfunctional. An irreducibly complex biological system, if there is such a thing, would be a powerful challenge

[6] Behe 1996, 69–73.
[7] Behe 1996, 39. Miller quotes this passage as well: Miller 2003, 294.

to Darwinian evolution. Since natural selection can only choose systems that are already working, then if a biological system cannot be produced gradually it would have to arise as an integrated unit, in one fell swoop, for natural selection to have anything to act on.

We can translate this passage to eliminate reference to 'functions': if a system is irreducibly complex with respect to some capacity x, then removing any part of it means eliminating the capacity of the system to do x. A system that cannot do x cannot be selected for by natural selection, so there is no way natural selection could build up an irreducibly complex system. The flagellum is clearly a biological system. If it is irreducibly complex, says Behe, then it cannot be the product of natural selection. Furthermore, he claims that no other known, unintelligent process such as symbiosis or self-criticality can account for the origin of irreducibly complex biological systems.[8] Therefore, biological systems which are irreducibly complex must be the product of intelligent design. A full reconstruction of this argument is presented diagrammatically in Figure 12.2.

This argument structure should look familiar – the top three boxes constitute the core of the argument from order, with 'irreducible complexity' filling in for 'order'. But the similarities to past arguments do not end there. The sub-argument supporting the claim that irreducible complexity is associated with intelligent agency should also remind you of an argument we met early on in Chapter 3. I am referring to the Stoic version of the argument from order. Behe's argument is nearly identical to that espoused by Balbus in Cicero's dialogue (see Figure 3.2), except that we have added natural selection and a handful of other mechanisms to the list of unintelligent causes along with chance. For this reason, Behe's argument is open to the same sort of objection as its Stoic predecessor: there is no reason to believe that the list of unintelligent causes is complete. In essence, Behe gives a negative argument. The idea is to eliminate a small set of plausible explanations (e.g. natural selection, chance, etc.) and from this negative result argue by default for intelligent agency. But there are few grounds on which to do so. The mere fact that Behe's list is longer than the Stoics' illustrates the point – by historical induction we should only expect more mechanisms to be added to the list of unintelligent causes. Thus, we can't be confident that we have eliminated all other possibilities.

[8] Behe 1996, 203.

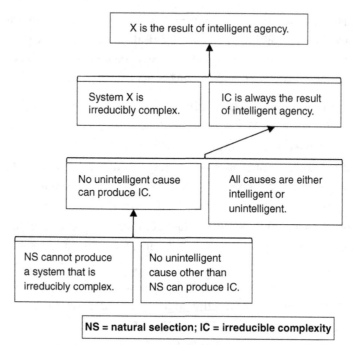

Figure 12.2   Behe's argument from irreducible complexity to the existence of a designer. 'NS' stands for Natural Selection; 'IC' stands for Irreducible Complexity

Whether or not this objection is decisive, there are other grounds on which to reject Behe's argument irrespective of whether all possibilities have been accounted for. Specifically, one could argue that in fact natural selection *can* produce a system that is irreducibly complex.

12.3 A counterargument: alternative capacities and evolutionary co-optation

Behe implicitly assumes that any given biological system possesses only one 'basic function' that can be favored by natural selection. However, when we naturalize the notion of 'function' in the most plausible way open to Behe, this claim becomes dubious. Biological systems generally have many capacities that directly bear on survival and reproduction and which therefore make the system a target of natural selection. It is this fact that Kenneth Miller accuses Behe of overlooking.[9] Because each biological system has

---

[9] Miller 2003.

many capacities relevant to the survival and reproduction of the organism, argues Miller, a system that is irreducibly complex relative to capacity x may be produced by natural selection initially favoring some other capacity y.

Miller's principal counterexample is the bacterial flagellum. As I mentioned above, flagella are very complex structures that are anchored in the cell membrane and which operate like tiny motors by converting chemical energy into rotational mechanical motion. The most prominent feature of a flagellum is the long whip-like structure which, when put in rotational motion, moves the bacterium around a fluid environment. Behe claims that systems which are irreducibly complex with respect to some capacity cannot be favored by selection when a part is removed. If he is right, then the flagellum – which loses the capacity to propel the bacterium when one of its component proteins is removed – cannot have intermediate structures that are favored by natural selection. However, Miller points out that a subset of the proteins in a typical flagellum make up a very important structure found in many disease-causing bacteria. That is, a flagellum missing some of its pieces is found to be highly favored by selection for an altogether different capacity than motility. The structure in question is the Type III injectisome which bacteria use to pierce the cell membranes of other organisms and inject a variety of proteins. An injectisome is really just a flagellum missing the long filament of the propeller and the proteins that bind the propeller to the drive-shaft.[10] So an intermediate system that contains most of the parts of a flagellum could have been favored by natural selection for its capacity to function as a syringe. To get a system favored for motility, only one or two additional proteins related to the filament would have to be introduced. So even though the flagellum is irreducibly complex with respect to the capacity for swimming, taking away one part does not give a 'nonfunctional' system – it gives a system with other capacities favored by natural selection.

Miller's example of the injectisome is a particularly simple version of a widespread process in evolution called *co-optation*. In evolutionary co-optation, systems evolved for one capacity are 'co-opted' to exploit another capacity, often in combination with other previously independent systems. To see how this might work, let's take a step back and think about just how

---

[10] For an excellent review showing in detail the structural similarities and differences between the flagellum and the injectisome, see Cornelis 2006, 39.

natural selection operates in Darwin's theory. Individual organisms are in constant competition with other members of their species and of other species for the resources necessary to survive and reproduce. Likewise, individuals struggle against the physical environment. Not all individuals in a given population can succeed and reproduce. Those that are successful will see their heritable traits increase in proportion in the next generation. This is natural selection in a nutshell. But note that I said *individual organisms* succeed or fail, not parts of organisms. While individual traits are heritable and can vary more or less independently of one another, it is the entire organism that competes and the entire organism that succeeds or fails. Thus, natural selection operates not on one trait at a time, but on combinations of traits. This means that, as far as natural selection is concerned, the relevant capacities of a system include all of the ways in which it contributes to the survival and reproduction of the whole organism in combination with every other system.

Now, it may be the case that some particular system – a set of feathers, for instance – has a capacity – insulation, say – that is favored by natural selection regardless of the status of most other systems in the organism. In other words, one capacity of a system can be so important to reproductive success that we can model its evolution by natural selection as a process involving just that capacity alone and ignoring the other systems in the organism. But other systems are nonetheless present, and are also being modified in a population over time. For example, the arm bones of the earliest dinosaurs to develop feathers were also subject to significant evolutionary change, as they lengthened for grasping and other uses we can currently only guess at. At some point, the feathers and the elongated arms in *combination* gave dinosaurs the capacity to glide. At that point, the most important capacities of each system so far as natural selection was concerned became those that allow for flight – each system became part of a larger system with the 'basic function' of flight. As the combined system was further refined by selection, the result was an irreducibly complex structure involving highly modified arms, flight feathers, specialized wing muscles, etc. In the finished product, the loss of any one part entails the loss of the most important, current capacity of the system, namely flight. But each of the component systems was evolved because it allowed the organism to do something else (e.g. grasp prey or keep warm). There is thus a perfectly plausible sequence of intermediate stages leading to a fully functional wing, each of which could be favored by selection. The point is that the most relevant capacity of any given system – its basic function – can shift in time as the context in which it is found changes.

12.4 Examples of co-optation

As I said, Miller's example of the injectisome is just a particularly simple example of co-optation. As Miller understands Behe, this one example is enough to defeat the biochemical design argument – since Behe asserts that removing *any* part from an irreducibly complex system renders it completely functionless and thus immune to selection, to prove Behe wrong it is sufficient to show that one such intermediate exists. However, Behe could object that demonstrating the possibility of one such intermediate is not enough. His claim is that natural selection cannot produce irreducible complexity and to produce a complex trait by natural selection involves an entire sequence of intermediates. So to prove him wrong, we would have to show that, at least in principle, an entire *sequence* of intermediates can be described for a given irreducibly complex structure, each of which would have been favored by natural selection. In the next section we'll look at just such a sequence – one that is not only plausible, but well supported by the available evidence.

The evolutionary sequence I have in mind terminates in the 'irreducibly complex' system that allows insects to fly. Flying insects, aside from the obvious specialized structures that are the wings, possess a suite of traits that make up a flight system. These include special, fast-twitch muscles for driving the wings, modifications of the integument that allow the wings to hinge and to be driven by big muscles which cannot directly attach to them, special neural structures that allow the rapid processing of information needed for flight control, etc. This system would appear to be irreducibly complex in that, if any of these components is removed, the function of flight is eliminated.

James Marden and his research group at the Pennsylvania State University have compiled evidence over the last few decades for a very promising theory of the evolution of insect flight, one that illustrates how co-optation can result in the irreducibly complex flight system.[11] The story

---

[11] For a recent review, see Marden 2003.

begins in the water. There is good evidence to believe that the common ancestor of both major groups of flying insects resembled modern stone-flies, a very 'primitive' group of winged insects known as the *Plecoptera*. Stoneflies lay their eggs underwater in streams, and the larvae are fully aquatic. They possess gills on their abdomens and gill plates along the thorax. Sometime in the middle of winter, the larvae pupate and the adult insects, which are not aquatic, emerge from the water, clinging to stones in the cold current. The adults mate on shore atop the snow (at least in North America). This means that they require a method of getting from the middle of a stream to the banks of the stream. And they need to do so in a hurry, since mates go quickly. There is wide variation in the methods and physical adaptations used by extant species to cross the water surface. Some essentially swim, some stand on the surface and use their wings as sails, some stand on the surface and use their wings like a fan-boat, and still others simply take to the air.

Marden saw this variety in modern locomotion, which can be arranged neatly in a continuum from skimming to flying, as a clue to the origin of insect flight. In Marden's hypothesis, which builds on work relating insects to crustaceans and showing insect wings to be homologous to gill plates, the first stage of co-optation occurred when adult forms of the insect began using their thoracic gill plates to push themselves along the surface of the water. At this early stage, only a few of the structures that would eventually comprise the flight system were present, namely the thoracic muscles that moved the gill plates. The plates themselves were selected largely for their capacity to circulate water over gills in order to oxygenate them. However, once this trait occurred in combination with the behavioral trait of surface swimming, the heritable combination was subject to selection for this new capacity. Swimming drove the develop-ment of longer gill plates, until eventually the insect could use them not just to paddle, but to sail. From passive sailing it was a short hop to active flapping, as species arose that could motor about on the surface while standing on their hind legs. It was at this point that a second co-optation is supposed to have occurred. Once the structures originally selected for usefulness in swimming at the surface were exposed primarily to the air by surface skimmers that stood in order to sail, natural selection could favor variations of this structure for aerodynamic rather than paddling properties. The paddles could become wings. Note that this did not have to happen 'all at once'. Even mediocre wings work well enough when they are being used to push a standing insect across the surface of the water. Gradual change in wing structure would be favored for every increase in aerodynamic efficiency, since there is a reward for getting to shore first (namely a mate). But no single mutation that yielded a wing was necessary – the intermediates all conferred a fitness advantage.

The point of this story – which as I have said is increasingly well supported by the fossil record, phylogenetic studies, comparative developmental work, and observations of extant species – is that, for at least one irreducibly complex structure, there is an entire sequence of intermediates each of which could be favored by natural selection. The point that Behe misses, and that Miller presses, is that the important capacities of intermediate structures or combinations of structures needn't be the same capacity natural selection favors in the final 'irreducibly complex' system. In fact, the capacities favored by selection in intermediates, such as the ability to shield gills, may not even be present in the final system. Examples like this abound in the literature of evolutionary biology, some merely hypothetical and untested, others well substantiated. Since it only takes one well-substantiated example to undermine Behe's argument, very few scientists or philosophers take irreducible complexity seriously as a mark of design.

One last point about co-optation bears emphasizing. Could this process really get us to an irreducibly complex system, sensitive to the removal of even a single part? Why isn't there redundancy in such systems even if co-optation is true? That is, why don't we see gills *and* wings on our flying insects if the wings are co-opted gill plates? The answer is best presented as an analogy borrowed from a wonderful little book by A. G. Cairns-Smith.[12] Cairns-Smith asks us to consider an arch of stones. Such an arch is analogous to an irreducibly complex biological system, in that if you remove any one stone (analogous to removing a component of the system) the whole arch collapses (ceases to function). Asking how an irreducibly complex system could come to be is like asking how one can build an arch. Certainly stacking the stones of the arch one at a time will not work. The answer is that you make a heap of stones and then remove the stones in the middle, leaving only those of the arch behind. In fact, that's how natural stone

---

[12] Cairns-Smith 1985.

arches are made through erosion. In the biological case, you build up the relevant components of an irreducibly complex system by selecting for other functions of these components. The important point to emphasize is that, once a combination of such traits has been selected for a novel capacity, any residual components involved in the original capacities are likely to be lost if those capacities are no longer required. So in our stonefly case, the adult stoneflies do not require gills since they have other structures for breathing in air. Thus, although gills were originally selected in combination with gill plates, once gill plates had been co-opted as sails or wings, the gills were no longer needed. In fact, they would be in the way and thus positively disadvantageous. Natural selection eliminates them, just as we pull the rocks out of the heap under the arch. It is in this way that we get clean, irreducibly complex systems with little or no hint of the roles the various components originally played in other systems.

12.5 Telling science from non-science: falsifiability and its detractors

As I mentioned at the outset of this chapter, ID arguments are often presented as conforming to the rules of empirical science. The question of whether ID is a 'scientific hypothesis' is irrelevant to whether any given ID argument is sound. Nonetheless, proponents of ID arguments often argue that ID explanations are more 'scientific' than those that invoke natural selection. This is taken to lend support to conclusions of intelligent design. It is implicitly recognized that science offers a sort of privileged route to knowledge about the empirical world. If one can demonstrate that a particular hypothesis is a scientific one, then an argument in favor of that hypothesis would carry all the weight of scientific justification. Critics of ID tend to focus even more than its defenders on whether ID is a scientific hypothesis. For instance, the National Academy of Sciences asserts that, "[c]reationism, intelligent design, and other claims of supernatural intervention in the origin of life or of species are not science because they are not testable by the methods of science."[13] It is therefore worth considering whether and to what extent ID is a scientific hypothesis. To do so, one must first account for the distinction between scientific and non-scientific

---

[13] Steering Committee on Science and Creationism 1999.

claims. This, however, is a very difficult and very subtle problem in the philosophy of science. The best we can do here is to give a cursory sketch of one major position and its attendant difficulties, and to consider how Behe's argument fares with respect to this criterion.

Since Behe and other ID proponents make much of 'falsifiability' as a criterion for distinguishing scientific theories from unscientific explanations,[14] this is the position we'll focus on here. It was developed and defended by the philosopher of science Karl Popper, who was interested in the so-called 'demarcation problem'. This is the problem of separating beliefs that are justified by 'genuine' scientific method from those that are not.[15] Popper saw the demarcation problem as the key to most problems in the philosophy of science.[16] The connection is clear – if you understand what makes scientific beliefs different from other beliefs, then you understand something about what characterizes scientific method. Popper advocated *falsifiability* as a solution to the problem. That is, he argued that falsifiability was a necessary and sufficient condition for a theory[17] to be scientific. He explained falsifiability this way: "[S]tatements or systems of statements, in order to be ranked as scientific, must be capable of conflicting with possible, or conceivable observations."[18]

Popper's reasons for focusing on falsifiability were logical. Suppose we have some theory T (a collection of propositions) that, if it were true, would mean that some other fact O (a proposition) about the world is also true. We can express this as $T \rightarrow O$ (read 'if T then O'). We might think that, if we subsequently observe O to be true, we have confirmed or supported our theory T. We could write this inference as follows:

$$T \rightarrow O$$

$$\frac{O}{T.}$$

But the above inference form is *invalid*. It is an instance of a fallacy known as 'affirming the consequent'. Suppose that T is just the claim that "All ravens are black," and O is the observation claim that "Raven X is black."

---

[14] See Behe 2003.   [15] Hansson 2008.   [16] Hansson 2008.
[17] By a theory I mean something like a collection of propositions about the world such as Newton's theory of mechanics and Darwin's theory of natural selection.
[18] Popper 2002, 51.

That is, suppose you observe some particular raven X to be black. It does not then follow that T is true since, for instance, some as yet unobserved raven might be of a different color.

Popper rejected accounts of the confirmation of scientific theories that he saw as committing this fallacy. Going out and 'verifying' the predictions of a theory lends no weight to the belief that the theory is true. Instead, Popper appealed to the following valid inference form:

$$T \rightarrow O$$
$$\frac{\neg O}{\neg T.}$$

The symbol ¬ means 'not'. What this argument says is that if O is false then necessarily T is false. In our raven example, suppose you observe raven Y to be white. Then it cannot be the case that all ravens are black, and thus we know that theory T is false. Popper saw the empirical testing of genuine scientific theories as consisting of attempts to prove the theory false, not to confirm it. Thus, only those theories that can in principle be shown to be false by observing their consequences count as scientific.

Most philosophers of science do not currently accept falsifiability as necessary for a theory to be scientific. The main reason is that Popper oversimplified the logical situation and ruled out essentially all of scientific theory in the process. Suppose we want to test a prediction of, say, Newtonian mechanics in an attempt to falsify the theory. In particular, let's suppose that we apply a force to an object by use of a spring with known properties and measure the acceleration of the object. If we find that the acceleration of the object fails to be proportional to the applied force and inversely proportional to its mass as Newton's laws require, have we then falsified Newtonian mechanics? Popper would have us believe so, but episodes like this occur all the time in laboratories around the world without the underlying theories being immediately discarded. Are scientists misguided or was Popper missing something? The latter seems to be the case. In my example, the predicted measurement of a particular acceleration is not a consequence of the theory alone. Lots and lots of background assumptions are also in play. For instance, we must assume that the object moves under more or less frictionless conditions, that our measuring instruments are functioning properly, that the mass has been measured accurately and in the relevant way, that all forces have been accounted for, etc. When testing a theory like Newtonian mechanics that has yielded so many accurate predictions, the first thing scientists are inclined to reject as false is one or more of these background assumptions, not the theory. That is, if we get the 'wrong' result, the first thing we check is the equipment, then the past measurements, then the assumptions underlying our particular experiment, etc.

The point is that real predictions for experimental outcomes follow from the conjunction of a theory T with many, many background assumptions A_i. We can represent the logical operator 'and' with the symbol '$\wedge$' and write this conjunction 'T $\wedge$ A_1 $\wedge$ A_2 $\wedge$ ... $\wedge$ A_n'. Thus, we have the following valid inference in the case of a negative experiment:

$$T \wedge A_1 \wedge A_2 \wedge ... \wedge A_n \rightarrow O$$
$$\underline{\neg O}$$
$$(\neg T) \vee (\neg A_1) \vee (\neg A_2) ... \vee (\neg A_n).$$

Here the symbol '$\vee$' means 'or'. To clarify, if we observe that the prediction O is false, we can only conclude that either our theory T or one or more background assumptions A_i is false. In this way, *none* of our physical theories are falsifiable in Popper's sense, and so it is dubious that falsifiability is necessary for a theory to be genuinely scientific.[19]

What does all this mean for Behe's design argument? Behe dwells on falsifiability largely for historical reasons. Much of the modern literature on ID is closely tied to a slightly older political and intellectual movement known as 'scientific creationism'. Scientific creationism attempted to cast various literal readings of the Bible in scientific terms so that the material could be taught in the public school classroom without violating the US Constitution's First Amendment. What's pertinent here is that scientific creationism was quite rightly attacked as pseudo-science – that is, as a body of beliefs not justified by the norms and methods of science but which nonetheless are presented as such. Unfortunately, one way in which this attack was advanced was in terms of Popper's falsifiability criterion.

---

[19] W. V. O. Quine (and P. Duhem before him) argued against the thesis that individual propositions could be verified or refuted one at a time – our scientific theories are woven together into a web that is tested as a whole. For a general overview, see section 2.1 of Ariew 2008. For Quine's argument, see Quine 1951.

Since Behe is anxious to distance himself from creationism, he is eager to show that ID is in fact falsifiable, and thus scientific. Doing so, as I suggested above, also grants ID the sort of epistemic weight that comes with scientific theories.

What follows for Behe if we grant that falsifiability is a sufficient condition for a claim to be scientific? Depending on how we interpret the term 'function', the claim that a particular system is irreducibly complex does appear to be falsifiable. We could simply remove a part and see if the system still functions. The claim that no unintelligent cause can produce an irreducibly complex system is thus also falsifiable. On these grounds we would be moved to admit that Behe's explanation of the features of certain biological systems in terms of intelligent design is in fact scientific. However, this amounts to little when considering whether the argument is sound. It is precisely because certain claims in his argument are falsifiable that Behe runs into trouble – it seems to be false that natural selection cannot produce irreducible complexity.

13 Intelligent design II: specified complexity

13.1 Overview

The second major modern design argument based on biology is attributable to a single person, William Dembski. Like Paley and Nieuwentyt and other proponents of the argument from order, Dembski thinks he has hit upon a general method of "design detection" that can sift "the effects of intelligence from material causes."[1] This detection of intelligence involves identifying a "mark of intelligence," or what previous authors have called a mark of design. He calls this identifying property "specified complexity," and, as we'll see, it amounts to a special sort of improbable event. Dembski claims that specified complexity succeeds where previous proposals for marks of design have not. After reviewing many uncontroversial cases in which we infer design, he was struck by the common use of what I've been calling the argument from order that appeals to a property involving small probabilities:[2]

> I noticed a certain type of inference that came up repeatedly. These were small probability arguments that, in the presence of a suitable pattern (i.e. specification), did not merely eliminate a single chance hypothesis but rather swept the field clear of chance hypotheses. What's more, having swept the field clear of chance hypotheses, these arguments inferred to a designing intelligence.

As our historical survey of design arguments made clear, identifying a property that allows us to eliminate all non-design hypotheses is not easy, and so Dembski's claim demands both attention and scrutiny. In the following four sections, I'll present the notion of specified complexity and consider whether it can function in a compelling argument from order.

[1] Dembski 2005, 2.   [2] Dembski 2004, 314.

This project is somewhat complicated by the many conflicting accounts Dembski has given of specified complexity. Though I will draw upon all of his major published accounts, I will emphasize two of them. The first is a textbook coauthored with Jonathan Wells and aimed at a general audience. It is entitled *The Design of Life*.[3] I have singled out this source since Dembski calls it "the definitive book on intelligent design."[4] The second source I'll use to reconstruct Dembski's argument is a more technical, unpublished report which is cited in *The Design of Life* as the most up-to-date statement of Dembski's theory of specified complexity.[5]

13.2 What is specified complexity?

As I said, Dembski is attempting to identify a mark of design, a property which a thing can have only as a result of some intelligent intervention in the history of the thing. Dembski calls his candidate for such a mark "specified complexity." This property involves three components: "complexity," "specification," and "probabilistic resources." In this section, I'll present each component informally, as Dembski often does. Let's start with the notion of complexity. For Dembski, complexity just means improbability. In common usage, complexity has something to do with structure and it isn't immediately obvious what this has to do with probability, high or low. To motivate his odd use of the term, Dembski asks us to consider a combination lock. The more combinations that are possible, the more parts the lock must have. So in a sense, the number of possible combinations – something like the number of states the lock can be in – is a plausible measure of the complexity of the lock. At the same time, the more combinations the lock has, the smaller the probability of stumbling across the correct one at random. So the improbability of an event – stumbling across the correct combination at random – is connected with the complexity of the lock.

What of "specification"? A specification is a special sort of pattern, and an event is 'specified' just if it is an instance of a specification of the right sort. What do I mean by a pattern? Dembski gives no formal definition, but he seems to have in mind any describable order or structure

---

[3] Dembski and Wells 2008.
[4] Dembski 2012.   [5] Dembski 2005.

to which an event can conform. That's rather abstract, so let's consider some examples. Dembski gives a number of them that have nothing to do with design in biology. Perhaps the simplest pertain to coin-flips. If we are talking about sequences of flips, then both "all heads" and "all tails" are patterns. Another example that recurs often in Dembski's expositions involves archery. An arrow striking a bull's-eye is a specified event, as is a series of arrows striking a lattice of bull's-eyes painted on the wall. Of course, not just any pattern will do for detecting design. After all, every event has some sort of describable structure and thus conforms to some kind of pattern. For this feature of an event to distinguish some events (e.g. those resulting from design) from others, a specification has to be more than merely a pattern; only some kinds of patterns are relevant. But events don't come with labels telling us which patterns they properly belong to. So how can we tell the difference between a 'cherry-picking' choice of pattern – one expressly chosen to guarantee a conclusion of design – and a legitimate specification? One way Dembski tries to narrow the possibilities is by requiring that a genuine specification be describable 'independent' of the event in question. By insisting on independence, Dembski is trying to exclude patterns that are rigged to precisely match an event. Such a pattern can always be found, and doesn't really help us rule out a chance origin for the event. Here's how Dembski illustrates the difference between an illicit pattern and a genuine specification in terms of the archery example:[6]

> Suppose each time an archer shoots an arrow at the wall, the archer paints a target around the arrow so that the arrow sits squarely in the bull's-eye. What can be concluded from this scenario? Absolutely nothing about the archer's ability as an archer. Yes, a pattern is being matched; but it is a pattern fixed only after the arrow has been shot. The pattern is thus purely *ad hoc*. But suppose instead the archer paints a target on the wall and then shoots at it. Suppose the archer shoots a hundred arrows, and each time hits a perfect bull's-eye. What can be concluded from this second scenario? Confronted with this occurrence, we are duty-bound to infer that here is a world-class archer, one whose shots cannot be attributed to luck but rather must be attributed to the archer's skill and mastery. Skill and mastery are, of course, instances of design.

[6] Dembski and Wells 2008, 172–73.

Painting targets around events after the fact is analogous to choosing inappropriate patterns – patterns that are *not* independently describable. On the other hand, describing the layout of bull's-eyes on a wall without reference to any arrow strikes (perhaps even before an arrow has been fired) is an appropriate specification.

Unfortunately, the notion of independence is ambiguous, and it is unclear that any of the possibilities can do what we need. Sometimes, as above, Dembski talks about independence as if it has to do with a logical relation between a statement of the event in question and the statement of a pattern to which it belongs. In this way, the pattern of bull's-eyes on the wall is independent of the event of the arrow striking a bull's-eye – treated as a particular point on the wall – because both can be described without reference to the other. But this way of posing independence forces us to account for the particular causal processes that bring about both the event and the pattern. We don't *have* to talk about the arrow to describe where the target is on the wall when the archer runs up and paints it. There is only a logical connection when we add extra statements explaining how the pattern comes to be realized. Only then is it true that the pattern entails the event, and so one depends logically on the other. But the whole point of Dembski's argument (and other arguments from order) is to detect design when we know nothing of the causal processes involved. Therefore, this notion of 'independence' doesn't get at the difference we need.

Sometimes Dembski treats the notion of independence as a matter of the order in which things are done in time: as long as you specify the pattern in advance of learning about the event in question, then the two are independent in the relevant way. However, that doesn't seem to help in the arrow example. Before I know where the arrow lands, I could specify the pattern "every arrow in a bull's-eye" even if that pattern is eventually realized by the archer running up and painting a target. Once again, we are only dissatisfied when we know something about the causal processes that brought about both the event and the pattern. There is nothing about the state of affairs with arrows sticking out of bull's-eyes that tells us something is amiss. What's really problematic about the case in which the archer cheats is that the probability distribution over outcomes is such that arrow strikes are guaranteed with probability 1 to conform to the pattern we've identified. It is this feature Dembski singles out in his third and final characterization of independence. That is, he often treats independence as a probabilistic condition. Specifically, he claims that independence is satisfied if the probability distribution over basic events is fixed without reference to the occurrence of any actual events. What this means practically, for Dembski, is that the distribution does not privilege any basic events – all basic events get the same probability. This guarantees that the probability of a basic event occurring (such as an arrow striking a particular point on the wall that is also a bull's-eye) is independent of whether the event conforms to the pattern of interest (e.g. that it is a bull's-eye). If E represents a basic event and S is the proposition that E conforms to a particular specification, we can write this in the notation of the probability calculus as $P(E) = P(E \mid S)$.[7] Of course, this condition also requires an even chance that any given event satisfies a given pattern. That's like saying that it's equally probable that any spot on the wall be a bull's-eye. To sum up, a specification is a pattern to which an event may conform and such that for a given pattern every event has an equal chance of conforming.

Dembski thinks he has identified a feature of patterns that is necessary and sufficient to guarantee that a pattern meets this condition of independence, and is thus a proper specification. This is the property of low 'descriptive complexity'. Descriptive complexity is, roughly, how hard it is to state or describe a given pattern. To use Dembski's own example, consider two sequences of coin-tosses: HHHHHHHHHH and HHTHTTTHTH.[8] The first is easy to describe to someone that hasn't seen it: "ten heads in a row." The second is much more difficult to describe without simply restating the whole sequence: "two heads, followed by a tails, followed by ..." In this sense, the first sequence has a much lower descriptive complexity. Low descriptive complexity is supposed to be indicative of intelligence.

---

[7] As I said at the outset, it's impossible to cover all of Dembski's myriad formulations in a single chapter. Perhaps his most thorough discussion of independence – which he calls "detachability" in this case – can be found in Dembski 2002. There he asserts that a rejection function f (which indicates those events too improbable to have occurred by chance according to chance hypothesis H) is detachable only if it is completely and uniquely determined by background knowledge K and $P(E \mid H) = P(E \mid H\&K)$. But if K determines f, it must determine at least part of the probability distribution used to assign probabilities to events such as $P(E \mid H\&K)$. Thus, $P(E \mid H\&K)$ is a meaningless expression – you cannot ask for the probability of the same probability distribution used to answer the question!

[8] Dembski and Wells 2008, 169.

At least descriptive complexity is a property of an event (or at least of our description of an event) and not a feature of the way we arrived at a choice of pattern. This, then, is the notion of specification I'll assume for the rest of the discussion: a specification is a pattern of low descriptive complexity.

The final component necessary for an event to possess specified complexity concerns what Dembski calls "probabilistic resources." These come in two varieties: 'replicational' and 'specificational resources'. The first is simply the number of opportunities available to witness a chance event. If I get to look at a thousand ten-flip sequences, it's pretty likely I'll spot one like the "all heads" sequence mentioned above. The thousand sequences I get to check constitute replicational resources. The number of random draws from an urn, the amount of time you allow a random process to run, the number of instances of a random process, the number of times you replicate an experiment – these are all examples of replicational resources. The more replicational resources you have, the more likely it is that the event could be the product of chance. For an event to exhibit specified complexity, it must still be improbable (i.e. complex) after all of the replicational resources have been factored in. This literally involves multiplication of a probability by the number of replicational resources.

Specificational resources are rather different. They "refer to the number of opportunities to specify an event."[9] To bring this into sharper focus, we might say that the specificational resources pertaining to an event are the number of equivalent specifications that the event could have satisfied. Equivalent here means of equal descriptive complexity. The idea seems to be that, while events matching specifications tend to suggest intelligence, we need to take into account how easy it is for an event to be declared 'specified'. If an event has a specification that is merely one of an enormous number of equally simple – and equally surprising – specifications, we want to adjust our assessment accordingly. So, for instance, suppose we go back to flipping coins and we come up with a run of ten alternating heads and tails, beginning with heads. That sequence is specified in that it fits a pattern with low descriptive complexity. But there are other specifications of equal complexity such as "a run of ten alternating heads and tails, beginning with tails" or "a sequence with five heads and

---

[9] Dembski and Wells 2008, 169.

five tails." There are also many of even lower descriptive complexity such as "a sequence with some heads." As with replicational resources, we want to factor out all of the equally plausible specifications. Again this involves a multiplication.

So what, then, is specified complexity? Informally, an event has specified complexity if it is an instance of a pattern with low descriptive complexity and is improbable even after factoring in the available replicational and specificational resources. When these conditions are satisfied, when something exhibits specified complexity, then we are entitled to infer design. One of Dembski's favorite examples is the bacterial flagellum on which we dwelt at length in the last chapter. According to Dembski, this structure exhibits specified complexity – it has so many parts in such a precise configuration that its chances of forming at random are vanishingly small. Furthermore, it conforms to a pattern of very low descriptive complexity: "a bidirectional motor driven propeller."[10] Because the flagellum possesses this property, we can, says Dembski, infer that it is the result of intelligent design.

13.3 Is specified complexity well posed?

Before we consider whether specified complexity is a reliable indicator of design, we need to try and clear up some difficulties with the definition. First, there is a bit of tension in Dembski's statements about specified complexity. I have uniformly referred to events, referring to both complexity and specification as properties of events. Confusingly, however, Dembski repeatedly speaks as if specified complexity can be asserted of a system, not an event. For instance, he asks "Are there any biological systems whose specified complexity is assertible?"[11] Likewise, he asserts that the bacterial flagellum, not its occurrence at a particular time and place, but the structure itself, exhibits specified complexity. But this is entirely senseless if complexity refers to improbability. The notion of a probability attaches only to events, not to objects. It makes no sense to ask for the probability of a coin, only the probability of the coin turning up heads when tossed or landing in the wishing-well when thrown. We will have to assume that Dembski is merely speaking loosely in order to avoid such awkward utterances as "the

---

[10] Dembski and Wells 2008, 174.   [11] Dembski 2004, 322.

occurrence of the bacterial flagellum at some time early in the Proterozoic." But because he often speaks only of systems, his key examples lack a clear presentation of the space of possible events being considered. This is a problem since according to his own recipe, we need to know the space of possibilities in order to decide whether an event is 'complex'.

Probabilistic complexity is ill-defined. The probability of an event given that it occurred by 'chance' is unknown or arbitrary if we have no statistical model of the process by which events are chosen. For instance, knowing that I've drawn a white marble from an urn doesn't tell me what the probability of such a draw is. I need a model of the drawing process. In particular, I need a model that tells me what the basic events are and how probabilities should be assigned to them. Suppose there are fifty black marbles and fifty white ones in the urn, the marbles are shaken before every draw, are drawn from the urn blindly, and are put back in the urn after they have been looked at. If the basic events are "drawing a white marble" and "drawing a black marble," then a good statistical model is one which puts a uniform probability over the two basic events. But if we don't return marbles to the urn after looking at them, our statistical model has to change. A good model is now dependent on the history of previous draws. In fact, it's easier to consider sequences of draws as the basic events. The point is, we have to know something about how an event came to be and what other events were possible in order to assign it a meaningful probability. Dembski claims that somehow the event itself specifies a full space of events and a unique 'chance' hypothesis, namely a uniform probability for every basic event. The first part of this claim is plainly false. A given event can be placed in an indefinite number of probability spaces. We must have a motivated way to select a space and an appropriate probability distribution pertaining to a relevant unintelligent, chance process. So far, this is just a worry that Dembski's account of complexity is incomplete. As we'll see, whether an event turns out to be complex depends enormously on how we fill in the event space and the probability distribution over events.

Specification is even more problematic. A suitable mark of design must be an objective property of an object or event, not a product of the perception or inclination of this or that observer. For these reasons, many published objections to Dembski's notion of specification center on the fact that it is a property of a particular description in a particular language of an event, rather than a property of the event itself. Dembski himself acknowledges this concern: "Doesn't our very choice of language bias us toward finding specified complexity (and therefore design) in certain phenomena as opposed to others? But in that case, how do we know that specified complexity is truly a marker of design and not merely an artifact of our choice of language?"[12] I would put the worry more starkly: as observers, we will designate as "specified" whatever events happen to be easily expressible in our language. But what is easy to say in English is not dictated in any straightforward way by properties of the events being described – descriptive complexity reflects only accidents of our language and does not correlate with any objective properties of objects or events. The concern is not that language biases us in one direction, but that descriptive complexity has nothing to do with designedness. In giving a rebuttal, Dembski interprets the objection as suggesting only that our choice of language can bias us to see design where there is none. After acknowledging the dependence of specified complexity on the observer's language, he says:[13]

> But this does not mean that the language we apply to a given range of phenomena was concocted on the basis of those phenomena and therefore biases us toward finding design in those phenomena when it actually isn't there. Consider the bacterial flagellum described as a "bidirectional motor driven propeller." That description came not from looking at molecular machines inside cells, but from human engineers who invented such machines before biologists ever suspected that they formed parts of cells. If anything, the ability of our language to efficiently describe the phenomena of molecular biology, far from suggesting that the specified complexity found there is an artifact of language, should lead us to think that the design in molecular biology is real. On this view, the language of engineering described molecular machines in biology because these are in fact engineered systems.

To make Dembski's counterargument sharper, we can write it this way:

(P1): If language causes us to declare design where there is none in the bacterial flagellum, it must be the case that the language was concocted with the details of flagellar structure in mind.

---

[12] Dembski and Wells 2008, 174.   [13] Dembski and Wells 2008, 175.

(P2): All of the terms in the specification of a flagellum as a "bidirectional motor driven propeller" were concocted without any knowledge of molecular biology.

(C): Therefore, our choice of language cannot cause us to declare design where there is none for the bacterial flagellum.

But this argument is plainly unsound because (P1) is false. There are many ways in which the choice of language can lead to spurious attributions of design given Dembski's scheme other than being influenced by the phenomena under consideration. For instance, one may simply lack the vocabulary to describe a kind of phenomenon in much detail. This is always the case for a new science. Think about the difference in the way you would describe an event such as the financial crash of 2008 and the way you would have described it when you were in grade school. I suspect that your grade school definition has much lower descriptive complexity, but only because you lacked the concepts and terminology to give a more thorough account. To look at it another way, if I told you that bacteria have a "bidirectional motor driven propeller" and asked you to draw it, I doubt very much that you would produce a schematic of the protein assemblage which drives the flagellum.

Irrespective of our limited vocabulary, words have varying degrees of abstractness. Choosing to describe the flagellum in terms of motors and propellers is to use a very abstract part of our language that permits shorter descriptions. We do stack the deck in favor of design if we use language of differing degrees of abstractness for describing the event (and thus fixing the probability space used to determine probabilistic complexity) versus describing the pattern (and thus determining descriptive complexity). For instance, if we described the event using the same abstract language Dembski uses to characterize the pattern to which it conforms, we might describe it as "bacterium appears with a motor" and then the obvious probability space has only two events: E_1 = "bacteria appears with a motor" and E_2 = "bacteria appears without a motor." Thus, the event E_1 should, according to Dembski's rules, have a probability of 0.5. It fails to be complex! On the other hand if we use the language of atomic arrangements for laying out the 'pattern' then the descriptive complexity of finding the bacterium in one such state goes way up. I won't try to write out an explicit description; perhaps it would be a set of coordinates for atoms of different types or perhaps, at a more abstract level, a description like "a configuration with approximate radial symmetry at the length scale of a nanometer with a high electronic density here and there and which tends to transition to states related by a rotation of ..." Either way, the space of basic events would presumably consist of all the states possible at this level of description – quite a large space, indeed. So we've recovered the intuition of high probabilistic complexity. But what about descriptive complexity of the pattern? Well, if we are consistent in the level of abstractness used in our language, the pattern also becomes much more complex. It's hard to say how one would characterize the notion of a motor in terms of atomic arrangements, but it would probably involve an enormous disjunction – this sort of arrangement or this sort or this sort ... The point is that we can and do bias the proceedings by failing to at least fix a common language for describing events and patterns.

For a starker illustration of the point, let's consider an event Dembski agrees is a matter of chance: a sequence of ten fair coin-flips. Suppose we have the genuinely random sequence given above: HHTHTTTHTH. This was supposed to have very high descriptive complexity. But let's introduce a new word into English: an 'orthosequence' is any sequence of ten coin-flips that contains roughly an even number of heads and tails. Now I can say that our sequence fits a trivially simple pattern: it's an orthosequence! By introducing some abstract terminology into a domain for which we have very little vocabulary already developed (namely coin-flipping) we're able to make descriptive complexity as small as we like. But this was supposed to be an obvious case where design is lacking. There's something wrong with Dembski's design filter.

To summarize the discussion to this point, the informal characterization of specified complexity as I've reconstructed it above is insufficiently developed to apply in an objective fashion. A great deal of essential detail is missing, and there is reason to believe that it cannot be made to work. Of course, one might reasonably worry that these shortcomings are artifacts of focusing on Dembski's informal treatment. However, as the next section demonstrates, we only find new problems when we consider Dembski's latest attempt at a sharp, formal, mathematical definition of specified complexity.

13.4 Specified complexity redux

In this section, I will present a streamlined version of Dembski's technical proposal. Wherever possible I will ignore unnecessary jargon, inessential definitions, and mathematical asides. Nonetheless, the reader who is uninterested in the nitty-gritty could profitably skip this section.

We begin with an event, presumably one exhibiting an interesting property that sparks our intuitions of design. In keeping with Dembski, we'll call the event E. This event conforms to a pattern captured by the description, T (which Dembski uses to stand for 'target'). To be as clear as possible, E is an event like getting heads when you flip a coin, while T is a syntactic object like the written expression "HHTHTTH" denoting a sequence of flips or the English expression "ten heads in a row." Dembski offers us two routes to eliminating chance. The first is cast in terms of Fisher's significance test. In general, a statistical significance test begins with a null hypothesis, such as "there is an equal number of black and white balls in the urn." For a sample of events – such as a sample of balls drawn randomly from the urn – one computes a measure of how far this sample deviates from what was expected according to the null hypothesis. One then determines the probability of getting a sample that deviates this much, assuming the null is true. If that probability is very low, then we can reject the null hypothesis. The details of Fisher's various significance tests do not concern us here, largely because what Dembski produces is not actually a 'significance test'. For a Fisher-style significance test, the 'events' in question are really samples – collections of data presumed to derive from the same distribution. For instance, if we wanted to test whether draws from an urn are random in the sense that there is an equal probability of drawing a ball of any color, then we would take a sample of, say, a hundred draws. This is the event, E, in Fisher's test – a collection of outcomes. Dembski, however, is not concerned with multiple draws. He generally treats E as a single event. Of course, in a sense we could transform Fisher's test into a probability distribution over sequences of draws of length 100. But it is essential that this distribution is entailed by the chance hypothesis distribution for single events. Since Dembski never speaks of samples, we may as well just skip the whole business about significance testing and just think about a probability distribution over the space of possible outcomes.

Taking this simpler view, here's how Dembski's proposal works. The event E belongs to a space of possible events (the basic events) that we'll denote Ω. Actually, E belongs to indeterminately many of these spaces, but Dembski seems to think of the choice of Ω as somehow given to us in the recognition of E. So, for instance, if I recognize E as a sequence of ten coin-flips, then Ω is supposed to be the space of all possible sequences of length 10. Once we've got an Ω, we're supposed to consider a special probability density over this space. Again, there are actually infinitely many possibilities here – for any given space there are uncountably many probability densities, and many of these plausibly correspond to non-intelligent processes. Dembski seems to think, however, that only one of these corresponds to the 'chance' hypothesis. That's the flat density that assigns all events (or regions of equal measure, but I'm glossing over such quibbles) the same probability.[14] Now, to eliminate the chance hypothesis, we pick some threshold probability below which we think the occurrence of the event by chance to be too improbable to be taken seriously. Call this number α. Now we check and see whether $P(E) < \alpha$. If so, we are supposed to reject chance and accept design. What happened to T? What role does specification play here? None, at least explicitly. Plausibly T plays some role, at least psychologically, in our choice of Ω, but Dembski doesn't tell us what role that is. All the work is done by the probability distribution corresponding to 'chance'.

In the end, this first version of specified complexity fails to give us the promised mark of design. It just says that we can eliminate one particular chance hypothesis (the flat distribution) for an event because it makes the event much too improbable. But we already know that such improbability is not by itself a sure sign of design. Boyle recognized centuries ago that non-intelligent processes can result in outcomes that are highly improbable on the crude assumption of a flat probability. And Bentley made it clear that a flat distribution over outcomes is not the right chance hypothesis to consider, even for the Epicurean. Rather, we should be looking at the distribution over outcomes assuming a chaotic initial state followed by law-governed evolution of the state. So this first version of Dembski's technical argument is a non-starter; we've seen much more compelling versions of the argument from order.

---

[14] Dembski consistently denotes this distribution by $P(* \mid H)$. But this is incorrect. A probability distribution for compound events (i.e. samples) on the assumption of a particular model (e.g. random draws) is not the same as a conditional probability. The latter assumes some larger probability space of propositions that includes the random-draw model. This latter propositional space of possible models is ill-defined and would push us towards the use of Bayesian methods rather than the frequentist techniques Dembski claims he is using.

However, Dembski offers a second formal approach, one that matches the informal account given above much more closely. In this approach, we begin by computing the probabilistic complexity of E as the probability that E occurs given some particular chance hypothesis. In particular, we assume that a given chance hypothesis H comes with a probability distribution $P(E_i)$ over a space Ω that includes E. As in our informal description of specified complexity, we thus begin with a probability, $P(E)$.

Now, the events we're looking for have a very low probability on the hypothesis of chance. But this probability we've got so far fails to factor in specificational and replicational resources. Replicational resources are easy to describe – they are simply the number of opportunities an event has to occur by chance. So, for instance, if we're wondering about a sequence of coin-flips, the replicational resources are just the number N of sequences sampled. If our sample is a million ten-flip sequences, then finding a sequence of all heads is not terribly surprising (or improbable). So as we build our measure of specified complexity, we want to factor in N. So that we don't have to insert a different value of N for every situation, Dembski suggests we use 10^{120}. That's a 1 followed by 120 zeroes. This is a very big number. Dembski chooses this number because, according to a somewhat dubious analysis by a computer scientist, it represents the total number of 'bit-operations' that could have occurred in the universe to this point in time.[15] We won't worry about whether asking about the bit-operations the universe could perform is a well-posed problem or whether the answer has anything to do with replicational resources. Let's just accept that there is almost certainly no case in which our replicational resources exceed this astonishingly vast number.

That leaves specificational resources. Here is where the pattern T comes in. T is a description of the event E in some language. According to Dembski, T has an objective degree of simplicity. He suggests that we equate the simplicity of T with something called Algorithmic Information Content or AIC. I'll have a lot more to say about AIC in the next chapter. For now, a sketch will suffice. The AIC of a string of symbols – like 0s and 1s in a binary message or the letters in this sentence – is the length of the shortest computer program that will reproduce the string when executed.

[15] Lloyd 2002.

Suppose we represent the AIC of description T by AIC(T). Then the specificational resources are given by:

$\varphi(T)$ = the number of descriptions (or patterns) T' such that $AIC(T') \leq AIC(T)$.

This is a sharp way of saying that the specificational resources are the number of simpler or equally simple patterns we might have attributed to T. As with replicational resources, we want to factor in the specificational resources. That is, we want to reduce the apparent improbability of the event by considering how many ways we might have specified it. Here is the final measure Dembski suggests as our mark of design:[16]

Specified Complexity: $\chi_H = -\log_2[10^{120} \, \varphi(T)P(E)]$.

The subscript H is supposed to indicate that χ is relative to a particular chance hypothesis and choice of probability distribution. I haven't said anything about why one would want to take the logarithm, but that has more to do with giving the measure of specified complexity some convenient properties. Imposing these properties is beside the point for us. What matters is whether χ is a reliable mark of design. There are really two questions here: (i) is χ a well-defined measure of anything, and (ii) is it a measure that lets us distinguish events caused by intelligence from other sorts of events? With regard to the first question, we still have most of the same problems we had with the informal version. In particular, it is entirely arbitrary what string one uses to represent a description T of the event E. Thus, while AIC is an objective measure of a property of strings, it is not a measure of an objective property of the event. But let's ignore these issues. Let's suppose that specified complexity as I've reconstructed it here is a perfectly acceptable property of events much like time and position. What follows for question (ii)? What does χ license us to say about design?

13.5 Specified complexity in context:
a failed argument from order

Let's suppose that specified complexity in one formulation or another is an objective, measurable property of events. Does the presence of specified

---

[16] Dembski 2005, 24.

complexity (or a high value of χ) indicate design? Dembski gives two main sorts of justification for believing that it does. The first is an eliminative induction. He claims that high specified complexity makes an event so improbable given a chance hypothesis H that it can be rejected with near certainty. Furthermore, we can quickly exhaust all known chance hypotheses, leaving only design. There are three problems with this argument. First, it is easy to show that high specified complexity does not succeed in ruling out some obvious non-intelligent processes. To give just one example, consider a crystal of ordinary table salt. If one were to peer closely at the crystal using, say, X-ray scattering techniques, one would see that the atoms of chlorine and sodium are arranged in accord with a very simple pattern. In particular they are laid out in a three-dimensional lattice. Whether one expresses this roughly with a descriptive phrase like "face-centered cubic lattice" or very precisely by describing the configuration of the unit cell and the dimensions of the whole crystal in terms of unit cells, the descriptive complexity will be quite low. It will certainly be too low to offset the minuscule probability of finding atoms arranged this way at random. That is, if we assume that the space of possibilities is any three-dimensional arrangement of the same number of atoms, there are an enormous (uncountable) number of such arrangements and so any one is vanishingly improbable. Even if we insist that the only relevant possibilities involve atoms in a lattice with the same spacing, there are still an enormous number of alternatives. So, under the assumption of uniform probability, the event we've witnessed is highly improbable and thus highly complex.

To put concrete numbers on things, suppose that there are only 10^6 atoms in our tiny crystal and there are thus 10^6 spots in a lattice to be filled. If every arrangement is equally probable (counting arrangements where atoms of the same type are permuted as the same arrangement) then there is a roughly 1 in 10^{301027} chance of getting the pattern we see. If the description of our lattice just listed the position of every atom in the unit cell and then told us the dimensions of the crystal in terms of unit cells, the description of the crystal, T, would be about 2,400 bits. In the worst-case scenario, AIC(T) is about the same as T. If we assume every possible program of the same length or shorter is a legitimate description of a pattern to which the crystal belongs, then $\varphi(T)$ would be of the order of 10^{722}. Even then, χ is about 10^6, far above Dembski's threshold of 1. In short, according to Dembski's account salt crystals must be the product of intelligent agency. But most people – Boyle included – would agree that salt-crystal formation is a paradigm instance of an unintelligent process. Something went wrong here – specified complexity is clearly not suffi- cient for design, and thus not a reliable indicator. This is not an isolated example. One could point to the fractal shapes of clouds and coastlines, the hexagonal flow patterns in boiling water, or the regular beats of pul- sars. In all of these cases, there is high specified complexity but no one, not even Dembski, thinks they are designed.

The second problem is that eliminative induction is always biased toward the default hypothesis. I raised this concern when discussing Boyle's approach to identifying final causes. Dembski wants us to assume that, having exhausted our imaginations with respect to unintelligent processes, an event must be the product of intelligent design. But if the survey of historical design arguments has taught us anything, it's that this is an unreliable strategy. The more natural science progresses, the more corrections we are forced to make. By insisting on eliminative induction, Dembski is taking a god-of-the-gaps approach, but history has shown this approach to lead to more or less continuous disappointment.

The third problem is that in claiming that specified complexity lets us eliminate all non-intelligent causes, Dembski is implicitly assuming that intelligence itself cannot be a consequence of physical (i.e. non-intelligent) causes. While this may be true, it requires an argument. There are many defenders of a view of mind contrary to Dembski's.

Dembski offers in alternate strategy for justifying specified complexity as a mark of design that appeals to simple induction:[17]

> The justification for this claim is a straightforward inductive
> generalization: in every instance where specified complexity obtains
> and where the underlying causal story is known (i.e., where we are not
> just dealing with circumstantial evidence but where, as it were, the
> video camera is running and any putative designer would be caught red-
> handed), it turns out that design is actually present.

Now, Dembski doesn't tell us what these known examples are, but the only uncontroversial instances of design come from humans. That is, the

[17] Dembski 2004, 320.

only objects we can all agree were designed are human artifacts. So let us grant Dembski that all such artifacts show specified complexity. What are we entitled to conclude from this generalization when approaching new objects about which we are uncertain? Well, if the new object is of the same sorts of materials and of the same size, etc. as some or all of the human artifacts we have examined, then if it displays specified complexity we may conclude it was designed *by a human*. On the other hand, if the object is nothing like the sample we examined – for instance, if it is an organism – then our induction does not allow us to infer anything at all!

This is really one of Hume's objections to the argument from order. In this case, since Dembski is appealing to straight induction, we don't need to accept all of Hume's epistemology to recognize the objection as compelling. Let me make the complaint clear with a simpler example. Suppose that all raven-shaped birds I have observed to date that have been black have also been found to possess a particular genetic marker X. Thus, I infer that within the class of objects consisting of raven-shaped birds, all of the black ones possess the relevant genetic markers. Now suppose I come across a big black bird with a raven shape. Then I can infer with reasonable confidence that it possesses the genetic marker X. Next, suppose that I come across a big black bird that is shaped differently (say, like a turkey-vulture). In that case, I am *not* entitled to infer that it will show the genetic markers – it does not belong to the class of raven-shaped objects on which my inductive inference was drawn. The point is, with straightforward enumerative induction we cannot go beyond the types of objects and causes which we have already observed to infer things about new types of objects. Thus, even if Dembski's specified complexity works for human artifacts, we cannot apply it to biological organisms, the objects of interest in his design argument.

As I suggested in Chapter 10, a successful argument from order will not appeal to such simplistic induction precisely because the controversial cases are quite different from the cases on which the induction is drawn. Rather, a successful design argument will appeal to something like a well-justified natural law that has as one of its consequences an association between intelligence and a particular property. Dembski provides no such well-justified law and no such connection between intelligence and specified complexity. In short, the argument from specified complexity is a sort of zombie argument, a version of the argument from order that died long ago but has been propped up with new technical language. Even so, I think we can grant Dembski the intuition that drives all arguments from order: certain properties, particularly those related to what we call 'complexity', elicit a sense of design. But what is complexity? Can we specify it carefully enough that we can make a compelling case for a mark of design? Could we perhaps show that only intelligent agency can result in complexity? We'll examine these questions in the next chapter.

14 What is complexity?

14.1 Life: a classic example of 'complexity'

The boundaries are hazy, but there does seem to be a genuine, objective distinction between living and non-living things. Unlike inanimate objects, living things reproduce, metabolize, and evolve. They are also vastly different in structure from non-living things. Living things tend to be highly organized at any given level of description and so differ from objects like stones, lakes, and volumes of gas which tend to have uniform or 'random' structures: Peer at a stone through a microscope and you'll find a jumble of randomly oriented mineral crystals. Look closer and you'll find the interiors of these crystals to be fairly uniform. Examine instead a cross-section of a sheep's eye and you'll discover intricate layers of tissue. Look closer at any one of these and you'll discover cells, each of which contains a visible nucleus. Use the right tools to look even closer and each of these cells will reveal myriad structures like mitochondria. You will discover fresh structure in the eye all the way down to the molecular level. So unlike a rock, the structure of an organism is highly layered. Like the rock, however, the structure of an organism at any one level of description is often asymmetric. Though from the outside humans are bilaterally symmetric, on the inside many of our organs – like the liver and pancreas – are not arranged according to any particular symmetry. Because of this asymmetry, living things are easily distinguished from such things as snowflakes, crystals, and man-made gears.

To illustrate the features of organisms I have been discussing, consider a typical dragonfly. Each dragonfly is approximately bilaterally symmetric at the highest level of description, at least in its external structure. But each of its parts is possessed of a subtle shape, not easily captured by the familiar constructs of geometry. For instance, the different curves defining the profile of the hind wings are shapes for which we human geometers lack a name. Furthermore, the closer we look at the dragonfly, the more structure we perceive, and the structure at one level – say the arrangement of simple lenses in one of its outsized compound eyes – is different from the structure at the next higher level of organization – for instance the placement of the eyes relative to the limbs. If we could peer closer and closer at this dragonfly, the amount of discernible structure would be enormous and varied. And, as with humans, some of that structure would appear asymmetric, such as the folding of the gut.

While the nested, asymmetric structure of dragonflies distinguishes them from non-living things like rocks and machine screws, these structural features are to some degree present in more elaborate human technologies. For instance, the sort of hierarchical and varied structure we see in the dragonfly is in many respects present in automobiles as well. From the outside, cars are bilaterally symmetric like dragonflies. But like a dragonfly, the closer we look, the more structure we see, where the symmetry and qualitative nature of that structure changes dramatically from level to level. Consider the engine of your favorite car. There you will still find a large degree of bilateral symmetry, but also a great deal of new, asymmetric structure in the form of hoses, clamps, cables, and gaskets that we couldn't discern by looking at the outside of the car. If we look closer at some of the engine parts, we'll discover much greater structure. For instance, in a high-performance engine, each spark plug is itself a small machine complete with an ignition coil that provides a spark to repeatedly ignite a fuel mixture inside the engine. The timing of this spark is controlled by an onboard computer that itself possesses many layers of structure with tiny, microscopic transistors at the very bottom level of description.

All this talk of structure is rather vague, but it emphasizes the sort of property needed to make the argument from order plausible. Recall that this argument requires as a premise the claim that some property – we've been calling it 'order' – is tightly associated with intelligent agency. If there is such a property, it ought to be shared by human artifacts – things we *know* were intentionally produced by intelligent agency – and those natural objects which we suspect of being artifacts produced by a more powerful intelligent agency. Often, the exquisite forms of living things are singled out as comprising this second group of objects. That is, to argue from the properties of organisms to the existence of a non-human designer of organisms, we need to identify an objective property – call it 'complexity' – that is shared by organisms *and* technological artifacts. The hierarchical, asymmetric structure that I've been talking about might fit the bill. So too might any number of features of the systems we call complex.

But what exactly is complexity? Is it possible to give a rigorous account of the notion that plausibly represents an objective property of objects rather than a subjective judgment of human observers? These questions, though motivated by altogether different goals, have been taken up by many researchers in fields as disparate as economics and ecology. The answers on offer are almost as numerous as those asking the question. Though I have focused on the notion of complexity as it pertains to structure, there are other aspects of objects and systems that seem to warrant the attribution of 'complex' such as the way in which parts at one level interact with one another and with parts at other levels. All of the different aspects of a system that elicit a sense of complexity are represented in Seth Lloyd's list of proposed complexity measures,[1] which is partially reproduced in Table 14.1 along with pertinent references. The complexity measures on this list are grouped according to three questions intuitively related to the complexity of a physical system: (1) how hard is it to describe the system? (2) how difficult is it to create such a system? and (3) how organized is the system?[2] Each proposed measure represents an attempt to define complexity by focusing on one of these questions. Any one of these definitions might just capture the sort of property we are seeking.

In the next section, I examine two of the more well-developed measures of structural complexity – measures that address how hard it is to describe an object or system – and identify a problem with applying these measures to physical systems. In section 14.3, I'll consider how other measures proposed to capture structural or other sorts of complexity might help us out of this difficulty. There are simply far too many proposals to do them all justice here, but we can nonetheless get a sense of how hard it is to pin down a slippery notion like complexity. At the very least, we can use

[1] Lloyd 2001.
[2] For a lucid overview of each of these aspects of complexity and a tour of the current state of 'complexity science', I recommend Melanie Mitchell's *Complexity* (2009).

Table 14.1 Measures of complexity

Difficulty of description	Difficulty of creation	Degree of organization
Algorithmic information content[a]	Logical depth[b]	Hierarchical complexity[c]
Minimum Description Length[d]	Thermodynamic depth[e]	Effective complexity[f]
Lempel–Ziv complexity[g]	–	True measure complexity[h]
Fractal dimension[i]	–	Fractal dimension[j]

Notes: [a] Chaitin 1966, 1975a, 1975b, 1977, 1979; Kolmogorov 1968; Solomonoff 1964a, 1964b.
[b] Bennett 1995.
[c] Simon 1962.
[d] Rissanen 1978.
[e] Lloyd and Pagels 1988.
[f] Gell-Mann 1995; Gell-Mann and Lloyd 2004.
[g] Lempel and Ziv 1976.
[h] Grassberger 1986.
[i] Mandelbrot 1983.
[j] Mandelbrot 1983.

a couple of concrete proposals to consider how the argument from order stands to profit from a development of the sciences of complexity.

14.2 Two accounts of complexity

One prominent measure of complexity is called Algorithmic Information Content (AIC). You will also see this or very similar definitions referred to as 'Kolmogorov complexity' and 'Minimum Description Length' (see Table 14.1 for references). AIC was developed in the context of information theory – it is not inherently a theory of the properties of physical objects. For this reason, I'll first have to describe how AIC works for abstract objects called 'strings', and then consider how we might extend the definition so that we can use AIC as a measure of the complexity of concrete physical systems like organisms or the solar system. As we'll see, this last step is the hard part.

To understand AIC, we have to begin with strings. A *string* is just an ordered list of characters like the list of numbers on your credit card or each of the sentences in this book. Information theory is a branch of mathematics that in part considers the properties of strings (like sentences) and the sources that produce them (like authors). In the context of information theory, AIC was developed as a means of characterizing the 'complexity' of a given string. The measure trades on the intuition that a complex string is harder to reproduce than a simple one – if a string is very complex, then intuitively I would have to give you a lot of instructions in order for you to make a copy of it. So the idea behind AIC is to use the number of instructions required in order to reproduce a given string as a quantitative measure of how complex that string is. Of course, we don't want to consider the sorts of instructions one human would give to another – these are much too informal. Instead, AIC is defined in terms of the sort of instructions one would have to give a computer in order to produce the string, namely a program. So very roughly, the AIC of any given string (usually assumed to be a binary array of 0s and 1s) is defined to be the length of the shortest program that will output this string when executed:

> **AIC**(x) =$_{df}$ the length of the shortest program[3] in some specified language that will output the string x.

So, for instance, consider the two strings x_1 = "011010100010100010100010 000010 ..." and x_2 = "100001101100100110000011100010 ..." The ellipses are supposed to indicate that each of these strings is very much longer than I have shown, say 1,000 digits long. Now the first 1,000-digit string x_1, though it may look random, can be easily produced by the following program:

```
for i = 1 to 1000

        if(is_prime(i))

        print 1;

        else print 0;

end
```

[3] This is with reference to a Universal Turing Machine, a Turing Machine that can 'simulate' any other Turing Machine. PCs are reasonable approximations of Universal Turing Machines.

In the language in which I have written it, our program is only fifty-eight characters long (including spaces) and thus $AIC(x_1) \leq 58$. Note that I can only say that the AIC of string x_1 is *no more than* 58. This is because there might be an even shorter program than the one I've written down that I have yet to discover. In point of fact, the AIC function is incomputable, meaning that there is no program that is guaranteed to return a value for the AIC of an arbitrary input string. So in this respect, the measure is not very practical. Note also that I have cheated a bit by writing a function 'is_prime'. It doesn't really matter whether I write out this function (thus increasing our AIC a bit) or take it as primitive component of whatever language we are using. Let's do the latter for now. So in summary, the 1,000-digit string x_1 turns out to have a low AIC because in essence there is a hidden pattern. All I have done is written a 1 in the nth place in the string if n is prime, and a 0 otherwise.

The second string x_2 is genuinely random. To make it, I flipped a coin a bunch of times and wrote down a 0 for heads and a 1 for tails. It turns out that, for a random string of this sort in which all characters appear with equal long-run frequency and any character is just as likely as any other to fill a spot in the string, the AIC is at least as long as the string. That is, the best we can do in this case is write a program that says:

```
print 10000110110010011000001110010 …
```

Of course, this example highlights one of the problems with AIC – it doesn't quite capture what we tend to think of as truly complex, even for strings. As Murray Gell-Mann points out,[4] the AIC measure would assign a random text typed out by monkeys a much higher complexity than it would the works of Shakespeare when the latter are considered as making up one long string. That is, the more *random* a string is, the higher its AIC, and thus AIC identifies complexity with randomness or disorder. While disorder of some sort may be related to our intuitive notion of complexity (think of the hoses in the automobile engine) it is certainly not everything. After all, we don't generally think that piles of sand are very complex, though they are quite random in their internal structure. For this reason, various modifications or alternatives to AIC have been considered.

---

[4] Gell-Mann 1995.

One such modification was proposed by Gell-Mann. He calls his measure 'Effective Complexity' (EC), and claims that it "corresponds most closely to what we mean by complexity in ordinary conversation and in most scientific discourse."[5] EC is built upon AIC and aims at fixing some of the problems with AIC. In particular, EC tries to compensate for the fact that AIC favors pure randomness. To Gell-Mann, it seems that when we constructed the AIC measure, we focused too much on how hard it is to describe the detailed form of a string, but forgot that not all of that form is 'structure' or 'order'. Complex things have lots and lots of structure, so our measure of complexity should be a measure of how hard it is to describe that *structure*, not how hard it is to describe the string itself. According to EC, then, we should compute the complexity of a string *x* by first accounting for its structure. We do this by describing all the regularities or patterns in the string. Our description of the regularities in *x* can in turn be written as a new string, *y*. Once we have this new string describing order, we simply find its AIC and use this number as an indication of just how much order there was in the first string. According to EC, this is what we should identify with the complexity of *x*.[6]

The intuitive idea of EC can be illustrated with a couple of simple examples. First, consider the grains of sand on a beach. The sand grains come in many sizes and shapes, and are distributed at random across the area of the beach. Were we to describe the beach with sets of coordinates, one for each grain of sand, the resulting string would look random and would possess a very high AIC. But beaches aren't very complex. From the EC perspective, we would look at our string of coordinates and note that there are no patterns. So our list of patterns would be very short, and thus we would assign the beach a low complexity. Next, consider the spaces in a parking lot. They are distributed in a regular array, and if we represented them as a string of coordinates as we did for the beach, there would be a couple of very obvious patterns. These patterns would encompass almost all of the data, and so our list of all the order in the string would again be very short (think also of my example with prime numbers above). Once again, parking lots have low EC. Finally, consider a high-performance automobile engine. If we again described the location and shape of every

[5] Gell-Mann and Lloyd 2004, 387.
[6] Obviously, I have given only a qualitative sense of how EC is calculated. For a rigorous treatment, see Gell-Mann and Lloyd 2004.

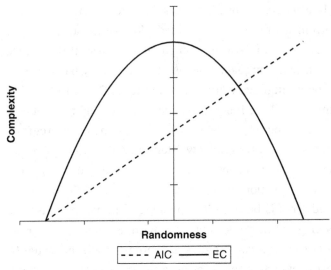

Figure 14.1   Comparison of AIC and EC measures of complexity

part in terms of coordinates, there would be a lot of little patterns evident
in the resulting string. After all, the hoses have a simple geometric shape,
but there are lots of hoses of different sizes in different orientations. The
spark plugs are arranged in a very regular pattern, but each has a great
deal of internal structure. The EC of an engine is quite high.

There are a lot of technical details I have been avoiding, but the simplest way to think of EC accurately is as a modification of AIC. AIC is a
function that takes a string x to an integer – it returns an integer 'complexity'. It behaves like the dashed line shown in Figure 14.1. The higher
the randomness of the input string, the greater is the AIC value of complexity. EC just adds an extra function into this process. We start with a
string x, apply a function to it to produce a new string y, then measure
the AIC of y. The function we add is selected to produce the behavior indicated by the solid curve in Figure 14.1. For EC, both very random and very
regular strings have low complexity, while strings that are somewhere in
between – strings with lots of highly varied order – come out with high
complexity.

We have so far been considering whether AIC or EC manages to capture our intuitive sense of complexity as it applies to strings. We have yet
to consider how either definition might be applied to physical systems –
what we are interested in is a property of things like rocks and rodents,

not abstractions like sequences of characters. We know that in the context of strings both measures are well defined. However, when we try to apply either the AIC or EC measure of complexity to physical things, we encounter a problem with language. The astute reader may have already begun worrying about language when I first defined AIC. Recall that AIC is defined in terms of the length of a program and one might reasonably expect the length of a program to depend upon one's choice of programming language. For instance, programs written in BASIC and in Java that produce the same outputs are generally very different in length. It turns out that this is not an insurmountable problem for either of the measures we have considered. This is because it is possible to prove the existence of a special class of programming languages with an astonishing property. For any one of these special languages, a program will be (roughly) as short as or shorter than an equivalent program in any other language. Furthermore, the AIC of any string relative to one language in this special class differs from the AIC relative to another in the class by at most a small constant. So as long as we're willing to tolerate a small degree of language-dependent sloppiness, AIC can be (almost) uniquely defined as the shortest program *in any language* that will output the string in question.[7] The problem with applying AIC to objects is thus not with the language used to construct the programs mentioned in the definition of AIC. Rather, the problem is with the language used to encode the features of a physical system as a string in the first place. What language you use to encode a concrete, physical thing like a table as a string makes an enormous difference for what the two complexity measures we have been considering say about the complexity of the system.

To be more concrete, consider just the AIC measure. If we want to use AIC to assess the complexity of a physical system, we must first produce a string that represents or describes that system. We would then determine

---

[7] More accurately, we can speak of programs as binary inputs to Turing machines that output binary strings. The AIC of string s relative to a Turing machine T, written $AIC_T(s)$, is just the length of the shortest input that has s as an output on T. Different Turing machines are like different programming languages. It turns out that there exists a Turing machine, U, such that for any string s and any other Turing machine T, $AIC_U(s) \leq AIC_T(s) + k$, where k is a positive constant that depends only on the choice of U and T. Furthermore, any two machines U and U' with this special property will disagree with one another on the AIC of any two strings by at most a small fixed value.

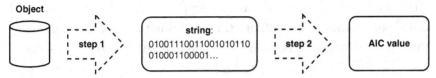

Figure 14.2   The process of computing an AIC value for a physical object

the AIC of the resulting string (see Figure 14.2). While the determination of AIC from a string (step 2 in Figure 14.2) is more or less uniquely determined, the string itself is not uniquely determined by the object. It depends on the language we use in describing the object (step 1 in Figure 14.2). If I describe a real object like a yellow rubber duck in a natural language like English, then the string description might be enormous compared to the description produced by taking a digital picture and making a string out of the pixel values – as they say, a picture is worth a thousand words. To make the point clearer, consider how difficult it would be to describe the rubber duck to someone in a language that lacked a word for yellow, and how trivial it would be in a language that had a word specifically for the sort of rubber duck you are talking about. How we choose to describe the object determines the nature of the string on which we will measure AIC. Thus, by changing the language in which we describe physical objects – the language we use to turn an object into a string – we can change which object is assigned a greater complexity. AIC does not seem to give us an *objective*, observer-independent property of complexity.

Unfortunately, because the definition of EC is built upon the definition of AIC, it too suffers from the same fundamental problem when applied to physical objects – the method or language we use to represent the object as a string determines which objects are more complex, ruining the objectivity of our measure. Now, that does not mean all is lost. There are a couple of ways of escaping this problem. First, one might argue that there is some inherently 'best' way to represent physical objects as character strings. That is, one might argue that there is only one function (or perhaps a few functions) for converting objects to strings that line up in some relevant way with the real world. For instance, if one is trying to represent objects in a manner appropriate for calculations in Newtonian mechanics, then the natural way to proceed involves specifying the relative positions and velocities of the centers of mass of each of the objects concerned. Perhaps there is some similar 'best way' one might argue for in the case of complexity attribution. Alternatively, one could attempt to devise a new measure of complexity distinct from AIC or EC that does not require us to represent the objects as strings. I'll consider this latter approach in the next section.

14.3 Other proposals

We know that AIC and EC suffer from a language problem – both measures depend on how we choose to encode objects as strings, and this is a subjective choice. But there are a variety of proposed measures of complexity that avoid this problem by making reference only to physical properties (as opposed to our *descriptions* of physical properties). One such example is 'thermodynamic depth' (TD). As I mentioned above, there are many aspects of physical systems that seem to be related to complexity. AIC and EC focus on how hard it is to describe a system. TD on the other hand considers how difficult it is to produce the system in question. To be concrete, let's take the system to be a particular snowflake. We'll also have to include (when considering TD) the volume of atmosphere from which the snowflake was created. Then, very roughly, we can say that the TD of this snowflake is roughly equivalent to the thermodynamic resources (e.g. internal energy, heat, mechanical work) that were expended to produce the snowflake from an initial state of air and water vapor according to the particular series of states the snowflake-plus-atmosphere system actually passed through. To determine the TD of the snowflake, then, we have to know that the flake began as a tiny hexagon of ice which then extended arms in each of six directions, etc. For each stage of the snowflake's known process of growth, we would then take stock of how hard it was to get to this stage from the starting point (in thermodynamic terms). If you like, you can think of the resources required to produce the snowflake at each stage from the preceding one as proportional to how improbable it was to arrive at the later stage from the earlier. This sounds very vague, but the idea can be made a lot more precise. I refer the interested reader to the references in the notes to this chapter. For our purposes here, this qualitative characterization will suffice.

The first thing to note about TD is that it is explicitly a function of physical properties of the snowflake – it does not depend on how we encode a description of the snowflake in a string. Thus, it is free from the difficulties faced by AIC and EC. However, actually trying to apply TD introduces a new set of problems. The first is practical: we generally do not know and cannot know all of the steps that went into the creation of the particular snowflake in question. We only have available to us a rough outline of how the process usually proceeds. This problem is much, much worse for such objects as bacterial cells or human beings for which many more steps are required. The second problem with TD is a problem in principle: the value we arrive at for TD will depend on which measurements we make of the process by which the snowflake came to be. If we measure different aspects of the snowflake (or measure the same things more frequently) we will get different estimates of how hard it is to make the snowflake. This does not mean that TD is a subjective measure of complexity – everyone making the same sorts of measurements would agree on a value. The problem is that we don't know how to decide which set of measurements to make. A problem similar to the encoding problem threatens TD unless we can find a way to stipulate what measurements should be made.

The upshot is that measures of complexity have been proposed that are 'physical' and so do not depend upon our choice of descriptive language.[8] Each of these measures has its appeal, and each has its drawbacks. Whether any is the right measure – or whether there is any one 'right' measure of complexity – is open for debate. But suppose we have one. What then? What could a convincing measure of complexity contribute to the argument from order?

14.4 Complexity and the argument from order

Recall that, for the argument from order to work, we need to identify a property of physical systems that is (almost) always the product of intelligent agency. Complexity in the loose, intuitive sense seems to be a property possessed by many human technological artifacts and by living things. Given the intense mental effort it takes for us to produce complex technology, this seems to be just the sort of property that might be associated with intelligent agency. If this is the case and if organisms also possess the same sort of complexity, then the argument from order would successfully lead us to conclude that there exists a non-human designer of living things. It is no accident, then, that the modern biological design

---

[8] The fractal dimension is another such measure (see Mandelbrot 1983).

arguments focus on complexity (witness Behe's 'irreducible complexity' and Dembski's 'specified complexity').

Of course, there are two major tasks to accomplish before an argument from complexity can be made. First, one must produce a rigorous, object-ive definition of complexity that captures the relevant intuitions, applies to physical objects, and succeeds in picking out technological and living systems as special. While physicists, mathematicians, and others inter-ested in complexity have yet to agree on a measure with all of these fea-tures, it is plausible that one or more can be found. Second, one needs to demonstrate that such complexity can only be produced by intelligent agency. Accomplishing this latter task appears much less hopeful. More and more, combinations of simple rules are shown to produce extraordin-ary complexity in both the intuitive sense and with respect to some of the formal measures defined above. For example, very simple iterative algo-rithms which involve the repeated application of just one rule are known to produce mathematical objects of extraordinary beauty, intricacy, and apparent complexity. Figure 14.3 shows a portion of the Mandelbrot Set. This binary map is generated by considering each point in the plane as a complex number, applying an operation to that number, then applying the same operation to the output, then applying the same operation to the new output, etc. Often, the outputs in this iterated process quickly diverge to infinitely large numbers. Each point in the plane is colored black or white depending on whether the process diverges when you begin with the number representing that point. Black is used to color points that don't diverge (the outputs stay finite as you repeat the process). The black points are the Mandelbrot Set. The point is that there is nothing especially intelligent about iterating a simple rule. One can write a trivial computer program to do it.[9] In fact, the one I wrote to generate Figure 14.3 is only five lines long. There is no intelligent agent involved in planning out the beautiful curves and intricate spirals of Figure 14.3. All of this apparent complexity is generated by a dumb process.

Similarly, the complex forms of many physical structures are now known to follow from simple physical laws. Examples abound in the litera-ture, but it is enough to point to the process of protein folding we discussed

---

[9] Similar algorithms which generate so-called 'fractal' forms are used to produce real-istic-looking landscapes in video games. See, e.g., Frade *et al.* 2008.

Figure 14.3   A portion of the Mandelbrot Set

in Chapter 1. Proteins can and often do have complex structures. But these complex structures, while difficult to predict, are known to arise from a process of chain folding that is governed largely by electrostatics. The amino acid chains themselves are produced by a complicated process also governed by relatively simple laws of electrostatics and quantum mechanics. We know (roughly) how these structures come to be and intelligent agency plays no role. But that means that this sort of structural complexity is not in itself a marker for intelligent agency. If it is true that complexity in the universe can arise from 'blind' physical law, then the argument from order appears to have stalled.

Of course, it might be tempting to suggest that it takes a very special combination of blind laws to produce complexity, and that a random choice of laws or of the physical constants populating those laws would be overwhelmingly unlikely to result in organisms. That is, one might attempt to argue that complexity is only the result of intelligence, and in some cases that intelligence works indirectly through the medium of physical law. But to argue this way is to make the sort of fine-tuning argument we will consider in Chapters 17 and 18.

15 Supernatural agents and the role of laws

15.1 Two questions

To this point in our survey of design arguments, we have been primarily concerned with proving that there exists a non-human designer of part or of the entirety of the universe. We have mostly considered the properties of the designer as a peripheral question to be answered after determining its existence. However, we are interested in the existence of God (or gods), not merely the existence of non-human designers, and God is supposed to be a supernatural agent. Often He is asserted to be omnipotent, omniscient, and perfectly good. In this chapter, we explicitly take up the question of whether and how one can infer the existence of a supernatural designer on the basis of empirical evidence. This, then, is the first question we want to focus on: can we infer supernatural agency? We'll consider one proposal for answering in the affirmative. Perhaps not surprisingly, the laws of nature figure prominently in the proposal, as they have in other design arguments we've considered. Here again is a central concern for natural theology to which we've paid only passing attention: what role do laws of nature play in design inferences? This is the second question that will engage us in this chapter.

15.2 A new vocabulary

In his book *Nature, Design, and Science*, Del Ratzsch attempts to provide some of the "foundational philosophical work essential for [the design debate] to make real progress."[1] Specifically, he sets out to clarify the concepts of 'design', 'pattern', and 'artifact' as well as the nature of mundane

---

[1] Ratzsch 2001, vii.

inferences to the existence of 'finite' designers. Lessons learned from the case of finite designers – especially human beings – are then used as a basis for showing how one might construct a strong inference to the existence of a supernatural designer.

Ratzsch lays the groundwork for his examination of design inferences by attempting to clarify and sharpen the definitions of key terms. Two principal definitions we'll need concern 'pattern' and 'design':[2]

(1) A *pattern* is an abstract structure which correlates in special ways to mind (it is a state of affairs which can be said to be 'mind-correlative').
(2) A *design* is a deliberately intended or produced pattern. Thus, we would say that to be *designed* is just to exemplify a design.

In the first of these definitions, it seems Ratzsch has attempted to clarify one vague notion – that of a pattern – with what appears on the face of it to be an equally vague idea, namely 'correlation to mind'. But this isn't being quite fair to Ratzsch. On the one hand, the intuition is apparent – he means to refer to whatever features of the world are comprehended or grasped by us, those ways the world is which happen to line up with our ability to contemplate or describe them. I'm not making the definition any more precise than Ratzsch does, but I am urging that there is a fairly clear intuition driving his attempt. At the same time, any attempt to really make this notion concrete must confront one of the most vexing problems in philosophy – the relationship between cognition and the world. Here is how Ratzsch puts it:[3]

> Identification of such [mind correlation] – that is, pattern noticing – presents itself to us experientially as a particular feel, a particular seeming, that defines our conviction that something makes sense, that we have gripped the correlation. The presence of this experiential dimension may explain why our talk in this area is so often metaphorically experiential – we "see" it, we "grasp" the matter, and so on. And we cannot get behind or underneath this experience to examine its credentials. Any evaluation of its credentials would have to employ resources and procedures whose justification would ultimately track back at least in part to that experiential dimension itself – the support for those credentials would have to strike us as themselves making sense.

[2] Ratzsch 2001, 3.   [3] Ratzsch 2001, 15.

So it seems that the best we can do in explaining the notion of a 'pattern' is just to say that it is whatever you mentally latch on to when you 'get it', when you understand something about the way the world is. Essentially any state of affairs you can describe is a pattern. The spiral arrangement of seeds in a sunflower is a pattern; the facts expressed by a set of blueprints for a building is a pattern; the bilateral symmetry of human bodies is a pattern, etc.

Aside from patterns, the central concept for characterizing design inferences is the notion of 'counterflow', which he defines as follows:[4]

(3) The term *counterflow* "refers to things running contrary to what, in the relevant sense, would (or might) have resulted or occurred had nature operated freely."

In this definition of counterflow, the term 'nature' is intended to contrast with 'artificial' in the sense of having been produced by an intelligent agent, and the term 'freely' means without the interference of intelligent agency. So, a process exhibits counterflow just if it would have worked out differently had only non-intelligent processes been in play. The existence of counterflow thus entails the activity of an intelligence. What any design argument attempts to do is establish counterflow.

One final definition will let us reconstruct Ratzsch's schema for design inferences:[5]

(4) An *artifact* is anything which embodies or exhibits counterflow.

So computers, houses, and kitchen utensils are all artifacts. But artifacts need not be designed. Something can exhibit counterflow even if the intelligent agent responsible did not intend to produce that thing. Ratzsch's example is a thoughtlessly whittled stick. For those of you with no experience of sitting on a porch mindlessly mutilating twigs with a knife, perhaps a more familiar example would be the regular pattern of footprints a person leaves behind on the beach. Artifacts are just those things produced by intelligent agents whether they mean to or not which would not otherwise be produced by non-intelligent processes.

The terms defined by Ratzsch stand in a rather complicated network of logical relations:[6]

---

[4] Ratzsch 2001, 5.   [5] Ratzsch 2001, 6.   [6] Ratzsch 2001, 6.

[P]attern entails neither finite design, intention, counterflow, agency, nor counterfactuality. With respect specifically to the finite realm, design does entail pattern, counterflow, intention, agency, and artifactuality. Artifact entails counterflow and agency, but not necessarily either intention or pattern (although it is obviously consistent with both). Counterflow entails artifactuality and agency, but neither pattern, design, nor intention.

The following chart summarizes these relations:

design entails ...	artifact, counterflow, pattern
artifactuality entails ...	counterflow
counterflow entails ...	artifactuality
pattern entails ...	none of the above

Ratzsch focuses on demonstrating the presence of design. However, all of the arguments we've been examining aim to establish the existence of a deity. It's only a secondary concern whether that deity intended to produce this or that feature of the universe. To show that a deity exists, it would be sufficient to show that some sufficiently grand portion of the universe – or perhaps the universe as a whole – is an artifact. According to the entailment relations above, it's enough to prove counterflow.

15.3 More about counterflow

Ratzsch views the unfolding of the world according to natural law as a three-step dance. It begins with 'initial conditions', the state of some part of the world at some particular time. It then moves through a 'process', and ends with a 'result', that is, a new state of the world. The notion of process he has in mind seems to be a chain of causal interactions: A causes B, B causes C, etc. There are thus three contexts in which we can spot counterflow. We might look at the current state of the world, which is presumably the result of a causal process from some earlier state, and immediately recognize this outcome as different from what nature acting freely would have produced. This is the sort of recognition claimed by Paley and Nieuwentyt upon stumbling across a watch in the field. By simply examining the watch we could be certain of counterflow, even in total ignorance of the process that produced the watch. But sometimes such immediate recognition based on static properties of the current state of the world is not possible. Consider any of our genetically modified crops containing genes for pest resistance. Scientists from another planet examining a sample of these plants would be unable to detect counterflow – there is nothing in their pest resistance or genetic structure alone that indicates that nature was thrown off course. However, if the alien agricultural scientists had been monitoring Earth during production of the crops, there would have been ample clues that nature was not acting freely in their production, clues like massive sequencing machines, strange seed production procedures, large expenditures of electricity, etc. Each of these things suggests that many of the steps in the causal chain essentially involve intelligent agency. Finally, it's conceivable that counterflow is evident in neither the result nor the process itself, but only in the initial conditions that set the whole thing rolling. Think back to the arguments of Newton and Bentley concerning the structure of the solar system. We have planets traveling in closed ellipses that are nearly circular. As both authors pointed out, the fraction of initial conditions – the proportion of ways in which the planets could be arranged prior to forming the universe – that would produce the result we see is minute. In other words, nothing in the arrangement of the planets themselves or the evolution of the solar system under the influence of gravity suggests counterflow. But the extraordinary initial conditions required for gravity to produce the system we have do suggest counterflow, at least according to Newton and Bentley.

But how do we actually recognize counterflow? What are the properties or features that tip us off in each of these cases? Ratzsch divides the marks of counterflow into two categories. The first he calls 'primary marks'. Chief among these are simple geometric properties such as straight edges, simple but precise geometric forms (like circles and pyramids), and regular spacing, all at a roughly human scale. "In houses, screens, fans, Stonehenge, cars, watches, gardens, and soccer balls, we see straight edges, uniform curves, repetitions, regularities, uniform spacing, symmetries, plane surfaces, and the like."[7] But that's not all. Primary marks include "uniformity of material (purified metal, glass, etc.), uniformity of color, uniformity of pattern (sometimes immaterial, as in algorithms), uniformity of sorting."[8]

---

[7] Ratzsch 2001, 9.   [8] Ratzsch 2001, 9.

Ratzsch gives us two reasons to believe that in fact primary marks tell us something about counterflow, representing two ways we might think about our knowledge of the normal flow of nature. The first is a straightforward inductive argument: in our collective experience, such properties are always associated with human agency. Of course, as Hume pointed out, we cannot use such an inductive scheme to argue for the existence of non-human designers, or in this case for counterflow arising from non-human intervention, because we lack the relevant experience. The second argument is quite similar to that offered by Cicero's Balbus: we appeal to our knowledge of natural law to assert that unintelligent causes simply do not produce precise geometric patterns on the relevant scale. Since states of affairs are necessarily the products of intelligent or unintelligent causes, it must be that precise geometric patterns are the product of intelligent agency. Of course, as with the argument in Cicero, the hard part is in justifying the claim that unintelligent causes cannot produce certain effects. Remember that the advent of Darwin's theory of natural selection added a new sort of 'unintelligent' cause to the list of known causes, and so dramatically changed our estimation of what can and cannot be produced by unintelligent causes. Ratzsch suggests that we simply take our current state of knowledge – the best scientific theories at hand – as constituting a full catalogue of unintelligent causes. For him, the relevant premise really says that no known, currently accepted physical theory suggests that precise geometric forms at the scale of 1 meter can be produced by natural processes. This is a sort of weak inductive inference, not a deduction. As he says, "we typically recognize artifactuality through recognizing indications of counterflow in results, processes, or initial conditions, and we recognize such counterflow against the background of and in contrast with our understanding of the normal flow of nature."[9]

So the primary marks are reliable indicators of counterflow because they are almost invariably associated with the actions of intelligent agents. These are to be contrasted with 'secondary marks'. These are characteristics that are frequently, but not invariably, associated with agency. They include the sorts of things usually pointed to in arguments from order: "complicated development, complex structures, coordination of components, adjustment of means to ends, interlocking functions, extreme

[9] Ratzsch 2001, 10.

improbability, purposelike behaviors, and so forth."[10] The association is loose. Secondary marks can occur in the absence of counterflow, and an artifact may fail to possess any secondary marks. Nonetheless, these marks are essential for making the inference to *design*. In Ratzsch's careful taxonomy, design involves both artifactuality and intention – an object is only designed if the agent that produced whatever pattern it exhibits intended to produce that pattern. According to Ratzsch, secondary marks are correlated with design, at least in instances where we know something is an artifact: "To take that step beyond mere artifactuality, we have to move from counterflow (involving things *nature wouldn't* do) to design (involving things *minds would* do). And although secondary marks do not provide the close connection to designedness that counterflow does to artifactuality, they frequently do constitute clues."[11] As we'll see shortly, secondary marks are essential for making the case for certain kinds of supernatural design for which primary marks are undetectable or non-existent.

15.4 Inferring finite agency

Suppose the Mars rover, *Curiosity*, comes across a titanium cube that is 1 meter long on each side, sitting upon the surface of the red planet. Such a cube has at least two salient properties from which we might argue for its artifactuality: its composition (the fact that it is pure titanium), and its shape. Let's just focus on its shape for now, but keep in mind that we could substitute other primary marks wherever I mention 'shape' or 'geometric form'. So what about the object's shape would lead us to believe that it is an artifact? Well, for starters, it's a cube. That is, I am supposing that the object has very sharp, very straight edges and smooth, planar surfaces such as humans might produce in a typical machine shop, and that the faces of the cube are rather precisely perpendicular, etc. The object has a very precise, very simple geometric form with a characteristic length of 1 meter. Now, every other object in our experience with a structure on the same length-scale that exhibits such geometric regularity is the product of intelligent agency, specifically human. That's why geometric form of this sort is a primary mark of counterflow. For this reason we can infer that the Martian cube is also an artifact. This argument is diagramed in

---

[10] Ratzsch 2001, 12.   [11] Ratzsch 2001, 13.

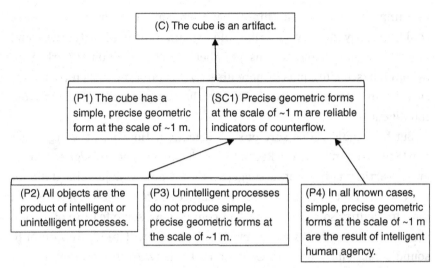

Figure 15.1   Argument supporting the claim that a one-meter-wide cube of titanium hypothetically discovered on Mars is an artifact

Figure 15.1. I have included both sub-arguments discussed above in favor of (SC1), the claim that the precise geometric forms are reliable indicators of counterflow.

It will be helpful to consider another example that, as we'll see, differs in some important respects. The example I have in mind is the real-life scientific program called SETI, which is short for "Search for Extraterrestrial Intelligence." Since 1959, scientists have been listening for signals from intelligent space aliens, principally using radio telescopes.[12] Readers may even have participated in this search through the "SETI@home" project, which harnesses idle processor time on home computers to process data from radio telescopes. The idea behind this search is that if such aliens exist, then their activity, whether intentional or not, might have resulted in the production of electromagnetic signals we can intercept and recognize as artifacts, or possibly even as messages to us. We've certainly spilled plenty of television and radio programming into the great beyond, and we've even sent some deliberate messages into space. And so we've been listening to see if extraterrestrials have done the same. Of course, we are assuming that we could distinguish deliberate, intelligently produced signals from natural ones. As Ratzsch points out, "The fundamental

---

[12] Sagan 1979.

presumption is that some things that aliens might produce, we humans could correctly identify, describe, and explain in terms of activities – and possibly intentions and designs – of aliens."[13] The astronomer Carl Sagan (whom Ratzsch cites) puts it more directly: "There is no difficulty in envisioning an interstellar radio message that unambiguously arises from intelligent life."[14]

But how could we be sure of such detection? How does the argument run? Sagan doesn't say, and Ratzsch doesn't explicitly go through the inference to artifactuality, but he suggests that it would look just like it did for the cube. Except in this case, rather than geometric simplicity, we would have to identify some other mark of counterflow, some feature of a radio signal that indicates it was not produced naturally. Here it's important to sound a cautionary note. In Chapter 11, I briefly described Jocelyn Bell's discovery of the first known pulsar. She had been monitoring a new radio telescope – precisely the instrument emphasized by SETI – and detected a remarkable, regular beat of radio pulses separated by 1.33 seconds. At the time, no natural process was known to produce this sort of regularity in electromagnetic radiation. So that regularity could reasonably be taken as a primary mark, and in fact Bell briefly toyed with the notion that she was looking at an alien signal. But it turned out to have a natural source after all: a rapidly spinning dwarf star with a massive magnetic field called a 'pulsar'. This episode is not in itself an objection to Ratzsch's argument schema for detecting artifactuality. After all, no primary mark is supposed to be a perfect guide to artifactuality – they all admit of some exceptions, and it looks as if this was one. We can cross this kind of regularity off our list, but there may be other marks, other properties of a radio signal, that are indicative of counterflow.

It is also conceivable, as Ratzsch seems to intimate, that we can make an argument by identifying design directly on the basis of secondary marks that are highly "mind-correlative." Consider the example Sagan gives of an unambiguous signal:[15] a series of pulses clustered together into groups containing 1, 2, 3, 5, 7, 11, 13, 17, 19, 23, 29, 31 ... pulses. That is, a train of pulses containing successive prime numbers. Now, it

---

[13] Ratzsch 2001, 18.   [14] Sagan 1979, 273.

[15] Sagan 1979, 273. This example also appears in his novel *Contact*, as well as in the movie of the same name.

is certainly true that we know of no astrophysical process that could produce such a signal. However, given the outcome of the pulsar case, it might seem hasty to suppose that no such natural process exists. But we might instead focus on the fact that the sequence is so profoundly resonant with human reason. It isn't exactly a message in English from the clouds, but it's pretty close. Prime numbers are significant to humans. That's precisely the sort of thing that we would transmit into space in order to announce our presence. Ratzsch suggests that the presupposition underlying Sagan's confidence is that the degree of "mind correlativeness" exhibited by such a sequence makes it a strong secondary mark. While secondary marks are not as strong an indicator of artifactuality as primary marks, they nonetheless constitute some degree of evidence. In the case of SETI, perhaps this is the best we can do. We have no idea whether such a strange signal could be the result of a natural process. But the strong secondary marks of regularity and complex structure that is of obvious significance to us provide some evidence of design, and, as a consequence, of the existence and activity of designers. This is precisely the strategy Ratzsch takes for certain kinds of supernatural design.

15.5 Inferring supernatural agency

As I said at the outset, Ratzsch's ultimate aim – and ours – is to explore the space of arguments that would allow us to conclude that a *supernatural* designer exists. Before we take up these arguments, we first need to be clear on what is meant by supernatural in the first place. Curiously, Ratzsch offers no explicit definition, but he does offer a number of characteristics of supernatural agency. A supernatural agent is in part an intelligent being with the following properties or abilities:

(1) It can use infinite or arbitrarily large quantities of any given resource (energy, matter, time, etc.).
(2) It might be able to violate or even change natural laws.
(3) Such a being might have much greater or perhaps unlimited knowledge of laws and the outcomes that can be achieved in accord with them.
(4) It might be able to create a cosmos and in so doing build patterns into initial conditions or the laws which govern its evolution in time.

Such an agent, says Ratzsch, could act in the world in one of four modes. The first, which he calls "nomic agency" ('nomos' is Greek for 'law'), involves a supernatural agent acting like a finite agent. There is nothing to stop a supernatural agent from producing outcomes by influencing initial conditions just like a finite agent could and allowing natural laws to play out as they would for us finite agents. This case is not terribly interesting from the perspective of natural theology, since there is clearly no way one could determine that an agent acting in this mode is supernatural rather than finite like us. The second mode, "supernatural nomic agency," is ostensibly different in this respect. Such activity would be structurally similar to regular old nomic agency: initial conditions, processes, and results would all be in accord with natural laws. However, the activity would require the employment of capacities beyond any finite agent. For instance, such agency might involve knowledge of natural laws or of initial conditions beyond any finite agent. Perhaps, for example, to pull off the intended result would require knowledge of the precise location of every particle in the universe at a particular instant. No finite agent could ever acquire this information let alone use it, particularly if the universe is infinite in spatial and material extent. The third mode in which a supernatural agent could act is called "supernatural contranomic agency." The term contranomic is a fancy way of saying "against the laws." Such activity involves going against the laws of nature in the sense that to achieve the result at least one law of nature must be violated or suspended. Finally, there is "supernatural creative agency." This involves activity that sets or determines the laws of nature themselves. The case of creative agency will be dealt with later. For now, let's turn our attention to the question of how one could infer the occurrence of supernatural nomic or contranomic agency.

Just as with finite agents, the strongest sort of argument for supernatural agency begins by identifying one or more primary marks of counterflow. So, for instance, we might peer through a brand new space telescope and spot a curious series of galaxy clusters arranged in a row with successive prime numbers in each cluster. That would be something! That sort of obvious 'regularity' would also be a primary mark of counterflow. From this fact alone, we could conclude that the galaxy cluster is an artifact. In the second stage of such an argument, we would have to demonstrate that the counterflow involved requires resources beyond those of any finite agent. In my galaxy example, it might be the case that the cluster series is so distant and thus so old that it is utterly implausible for any biological agents to have developed prior to the formation of the clusters. Or perhaps it could be shown that an effectively infinite amount of energy was required to arrange the clusters. This would lead us to conclude supernatural nomic agency. Most striking of all is the case in which we are able to show that the counterflow that resulted in the artifact involved the violation of natural law. Suppose, as is entirely likely for an ancient series of galaxy clusters, that tweaking the primordial mater distribution of the universe so as to produce the cluster as the universe evolved would require simultaneous coordinated action over an impossibly large spatial distance. That is, rigging the matter field so that the special series of clusters would develop would require communicating force at greater-than-light speed. From this fact, we could conclude that not only is the strange string of galaxy clusters an artifact, but that it is an instance of supernatural contranomic agency.

What if there are no primary marks? Can we discern design or artifactuality in instances in which there is no detectable counterflow? Absent primary marks of counterflow, we're right back where arguments from order usually begin – looking for "secondary marks" that are tightly correlated with intelligent activity. Ratzsch offers two proposals. The first is that we might identify an instance of design by virtue of its "mind affinity."[16] Mind affinity is another term for mind correlativeness, and neither is very precise. To illustrate the idea, Ratzsch presents three scenarios, supposedly in increasing order of mind affinity. Imagine that when we first mapped the far side of the Moon, we uncovered the crater field that is actually there. The craters are randomly arranged and distributed in size according to a simple probabilistic distribution. This first case – the actual one – exhibits very low mind affinity. Now consider a second case: the craters on the far side of the Moon turn out to be arranged in a more or less perfect grid. In this case, there is great mind affinity in the regularity of the phenomenon. Grids are arrangements that people just "get." We notice them quite easily and even like to arrange things in grids ourselves. In this case, though there is no evidence of counterflow – the grid might well be a product of natural processes – we are nonetheless inclined to

16 Ratzsch 2001, 63–64.

conclude design. Finally, in the third case we imagine that the craters are so arranged as clearly to spell out "John 3:16."

Here we might reasonably object that this is more about familiarity than affinity. We are used to humans producing such patterns as grids, and humans are the only things we know of that produce text. So strong is that association that we immediately infer a mind just like ours when we see English text. But as an argument, this is at best the sort of simple induction Hume criticized. Whatever we might conclude were there a biblical text written on the face of the Moon, none of the genuine cases under consideration are like that. Insofar as they differ from the effects humans produce, the inference to an intelligent cause is weakened to that degree. Of course, in the case of craters forming text, we might instead try to appeal to language use as uniquely the product of mind (though a static message is weak evidence of language use). But this strategy also runs into difficulties, as we saw in Berkeley's case back in Chapter 6. So the proposal that we appeal to mind affinity is too vague and ill-justified to be helpful, and insofar as it can be made precise, it just takes the form of one or another of the arguments we've already examined.

Ratzsch's second proposal involves value: "Secondary marks instrumental in production and preservation of value can in conjunction with that value constitute evidence of design."[17] Recall that secondary marks are features like complexity or extreme improbability that, when present along with primary marks, strongly suggest design as well as artifactuality. But, says Ratzsch, they can do the job alone as long as the presence of the secondary marks is essential for guaranteeing the presence or emergence of something with inherent value. What does he mean by inherent value? There are lots of things – life, beauty, love – that we humans think are valuable in and of themselves. These things need not serve some particular end – they need not have instrumental value – in order to be worthwhile. The pursuit of beauty, say, needs no justification. Beauty is an end in itself. There may be cases in which such ends are secured by processes or states of affairs involving secondary marks. So, for example, if it turns out that the emergence of intelligent life on Earth was the product of an enormously improbable event in evolutionary history, like a rock crushing just the predator at just the right time, then we might suspect that

---

[17] Ratzsch 2001, 70.

the emergence of intelligent life is a designed event. This is essentially the argument from providence – the secondary marks suggest the world was set up just right to result in this or that valuable end. In the argument from providence, the end is the benefit or happiness of humankind. The problem with such an argument, of course, is identifying inherent value. Sure, happiness is valuable to us, but how do we know it is an end in itself? More to the point, how do we know it is an end that a supernatural agent would set for itself? The problem of determining whether an outcome – valuable or not – is a plausible end is the problem of detecting final causes, and I refer you to the discussion of Boyle in Chapter 5 and Janet in Chapter 10 for an examination of the difficulties inherent in attempts to answer that question. The way Ratzsch frames it, we need to identify not merely plausible ends for this or that sort of agent, but the existence of objective values that all intelligent agents would strive for. The existence of such values is debatable.

15.6 Reformulating counterflow and the inference to finite agents

What Ratzsch offers that we haven't seen in previous versions of the argument from order is the notion of counterflow. As we saw above, this notion is also essential to providing a plausible argument for the existence of a supernatural agent. But as it is presented by Ratzsch, the notion is problematic, or at least incomplete. The problem is that pretty much everything exhibits counterflow. More specifically, every cause C of an event E exhibits counterflow: the presence of C results in an outcome different from that which nature would have produced had it operated freely, that is, without C. Perhaps Ratzsch means by "freely" something like "without the intervention of a free will." But this doesn't help much; we'd have to know exactly what a free will is (not an easy question), whether any such thing exists (not an easy question), and how its effects can be distinguished from those of natural causes. Remember, in the case of finite agency, no laws are supposed to be broken. In other words, Ratzsch wants to allow for the possibility that intelligent agents are themselves products of natural processes. So we don't want to suggest that free wills are unbound by laws.

We can sharpen up the notion of counterflow. To introduce the revisions I'm advocating, it will help to consider an example. Think back to

John Maynard Smith inspecting captured German warehouses in the Second World War. There is general agreement that he and his colleagues were acting rationally when they concluded that various hunks of machinery whose purposes were entirely unknown to them were nonetheless artifacts and likely designed ones at that. This is for Ratzsch a simple example of detecting primary marks of counterflow: geometric simplicity in the shapes of machine parts or cowls, material uniformities, etc. But let's see if we can't make this more precise. So as not to put words in Maynard Smith's mouth, I'll imagine that I'm the one inspecting the warehouse. When I look at an object in the warehouse, I draw upon many different regularities or natural laws to assess its status as junk or artifact. For instance, suppose that the object has large parts made of a metal that is highly uniform and is an alloy. Chemistry tells me what it takes to get such an alloy, and it requires a rather specific combination of pure materials. Geology tells me that such combinations are just never found in natural rock formations, let alone in configurations where they might pass naturally through the necessary temperature cycle to make the alloy. There is more. Physics tells me something about what kinds of initial conditions would result in the overall shape of the large metal parts as they crystallize from a molten state. Again, field observations in geology tell me how rare these conditions are outside of human factories. The same goes for most of the parts associated together in the putative machine. So, the conditions necessary to produce the thing according to the laws and known regularities of nature just don't seem to happen unless they involve human agency. No one of these cases requiring rare initial conditions may be compelling – I've seen some remarkably cubic crystals, for instance. But in conjunction they carry great weight. To accept that the item under examination was produced without human intervention would require me either to reject established laws in a slew of sciences, or else to believe in the occurrence of a confluence of initial conditions that all of human experience tells me just never occurs.

Note that I have to consider each putative artifact I find on a case-by-case basis. There is nothing inherent in any of the features Ratzsch identifies as primary marks that ensures artifactuality. Rather, we need to consider one property at a time and see if we can spot a combination of initial conditions and natural laws that would result in that property. If the only such combination or combinations involve initial conditions that just don't seem to happen unless people set them up, then we can conclude that the object is likely a human artifact. Otherwise, we remain uncertain. I'm suggesting that Ratzsch is right in that we recognize design only against a background of regularity or natural laws. When I infer that the strange object in a Nazi warehouse is a human-designed machine, I have to appeal to the laws with which I am familiar as well as to my knowledge of 'natural' and 'artificial' initial conditions, where the latter refer to conditions humans can and do produce. But I am also suggesting that there is no general answer to what nature would do if it acted 'freely', that is, without intelligent agents getting involved. Yes, simple geometric forms at the scale of 1 meter are, in our experience, always man-made. But for inferring the existence of non-human artifacts – whether made by aliens or gods – this simple induction isn't good enough. That's why Ratzsch appeals to counterflow. But the only way to make counterflow precise enough to do any work is to appeal to a broader induction over the kinds of initial conditions that tend to be realized in nature absent human intervention. Rather than attempt to discern what nature would or could do, we appeal only to what we know nature generally doesn't do without people, and this appeal is made with respect to initial conditions.

It is interesting that the kinds of features of the world relevant to design inferences are of a sort not directly covered by any scientific law. For instance, there is no law of medium-sized objects of mixed elemental composition. There is no universal generalization that tells me whether things like blenders can form geologically. This is different from the way in which we usually think of laws ruling things out. For instance, Newton's laws guarantee that a pendulum cannot swing higher on its return than the point from which it was released. Positions and energies of masses are things covered by Newtonian laws. But when we ask about whether a blender is an artifact, we're not asking about it as a mere mass or object in motion. We mean to refer to a detailed collection of properties – its mass, its shape, its material composition, etc. These complex bundles of properties are not jointly covered by any one law. Instead, scientists account for the origin and function of such systems by building complicated, often one-of-a-kind models that draw upon many separate scientific domains and which are full of approximations and idealizations. Inferences drawn from these models are a lot shakier than inferences drawn straight from the laws themselves. The only way to draw an inference of artifactuality for any given object is to use such a complex model to determine, so far as possible, how many different sets of initial conditions would result in the object in question, and to assess whether any of the possible initial conditions are naturally occurring. Of course, for everyday human artifacts, we usually don't bother to go to all this trouble. But everyday inferences based on simple induction do not hold up in cases of design by a putative extraterrestrial or god.

15.7 Problems for the supernatural inference?

Whether I'm right about how best to conceive of counterflow and the ordinary inferences to finite design that notion supports, the strongest inference to supernatural design rests on 'contranomicity' – the violation of natural law. The question is whether we are ever in a position to know that a law of nature has been violated. Before I explain what I mean by this question, let me explain what I don't mean. I am not asking whether it is 'scientific' to talk about the violation of laws or to hypothesize intelligent agents that can contravene the laws. For the project we are pursuing here, it makes no difference whether a hypothesis is scientific or not. What matters is whether it is rationally justifiable on the basis of empirical facts. Like Elliott Sober, I think that claims about the existence of deities can be crafted so as to function as scientific hypotheses.[18] Like Ratzsch, I agree that trying to draw a line between inferences to the existence of intelligent extraterrestrials as SETI attempts to do and inferences to God is ill-motivated and probably hopeless. The problem I'm pointing to is not about what makes for an acceptable scientific hypothesis, but rather an epistemic problem: are we ever justified on empirical grounds in claiming that a law has been violated?

It seems not. Science operates under the assumption that there are laws, or at least broadly applicable regularities, behind the phenomena of the world. Laws are confirmed or refuted on the basis of observation. If an observation fails to conform to a law then there are three conclusions one might draw. First, that the report of the observation in question is simply false – something was wrong with the experiment, there was too much noise, or some other factor was present that prevented us from

---

[18] Sober 2000, section 2.7.

getting reliable data. Second, the observation is correct and the supposed law is simply incorrect. Finally, we might conclude that the law is correct and was temporarily contravened by a supernatural agent. I suggest that there are two attitudes or rules for scientific reasoning one might adopt with respect to these options. On the one hand, one might insist that it is never acceptable to conclude that a law was violated – one should always seek a better law that would accommodate the aberrant observation. On the other hand, one could hold that sometimes, under certain conditions, we should accept that a law was violated and conclude that a supernatural agent was active, not that the law is wrong. Suppose nature is governed by laws – there is a set of laws that in fact accounts for all empirical phenomena – and supernatural agents never intervene. If we adopt the first attitude and keep looking for laws when we spot violations, then no matter what the laws are, we'll settle on the correct laws in the long run if they are discoverable at all. On the other hand, the second strategy makes us very easy to fool. There are many ways nature could be such that even if the laws are discoverable, we wouldn't learn them. Instead, we would quickly conclude supernatural agency and give up looking for better laws. So as long as supernatural agents don't contravene the laws, we're better off with the first strategy. What if supernatural agents do from time to time contravene otherwise inviolable laws? If what we mean by having them contravene the laws is that there simply are not any generalizations that encompass all of the empirically accessible facts, then someone following the first strategy would still be right in the long run – she would never settle on a complete set of laws covering every event, and so some events would remain outside the laws. Someone following the second strategy could still be tricked into attributing supernatural agency to the wrong events because she would discontinue the search for better laws too soon. In short, the best strategy for discovering laws and their violations is the first, in which we always look for a more comprehensive law. However, this strategy works only in the long run. At no point are we ever entitled to declare that these are the laws and these are the exceptions. All of our conclusions are tentative at any finite time in the future. Thus, we can never conclude that an apparent exception is in fact an instance of contranomicity. As a consequence, we cannot be justified in inferring supernatural agency given any finite quantity of empirical data.

15.8 The third case: tinkering with the laws

I have put off a discussion of Ratzsch's notion of creative agency. This is one of the possibilities he holds out for supernatural agents. A being capable of creating the cosmos may leave evidence of this creation in the details of the initial conditions of the universe or in the laws of nature themselves. That is, a supernatural designer may have built the cosmos in such a way that we can detect signs of design in the way laws were tinkered with or initial conditions rigged. For initial conditions, the intuition is that we can spot design if any universe with the laws we find in our would have to be started in one of a very restricted set of initial states if intelligent life (or some other inherent good) is to develop. There are echoes here of the ancient intuitions of conspiracy. Likewise, we might come to find that the laws of nature are set up such that if they were only slightly different the universe would appear vastly different from how it is and would not support intelligent life. Both of these options are taken up in detail in Chapters 17 and 18. In the meantime, let's turn to a closer look at the laws themselves.

16   A brief survey of physical law

16.1 Viewing natural laws as artifacts

Most of the design arguments we've encountered focus on parts of the universe and argue that this or that part of the world could only have come to be the way it is because some intelligent agent helped make it that way. As we turn to consider the phenomena of so-called 'cosmic fine tuning', we will find that the focus shifts to the universe as a whole. Rather than worry about properties of things in the world, a fine-tuning argument for the existence of God emphasizes properties of the universe. Typically, the proponent of such an argument looks to the form of the particular physical laws thought to govern the unfolding of the universe, and in particular to a set of 'constants' that feature in these laws. The relevant intuition is that the laws would not have the form they do were it not for the will of an intelligent agent. In a sense, fine-tuning arguments attempt to establish that the very laws which shape and govern the universe are artifacts.

In order to assess the plausibility of fine-tuning arguments, it will help to have some idea of what is meant by a 'physical law' and what laws in particular are currently thought to govern our universe. This chapter presents a survey of the major physical laws invoked by design arguments. I won't attempt here to give a philosophical characterization of what a law is – we'll worry about that later. Rather, my aim is to present the laws that will serve in key premises in the next two chapters, and to do so in a manner accessible to the non-specialist. The reader who is comfortable with the basics of modern physics can profitably skip this chapter. Similarly, the reader who wishes to avoid all technicalities can leap ahead to Chapter 17 and still be able to follow the gist if not the details of the fine-tuning arguments. For everyone else, I offer this guide to modern physics in three short acts.

16.2 A survey of natural law: the world at 1 meter

Physical theory – the set of laws posited to account for the behavior of matter in space in time – is immensely complex and highly interrelated. Nonetheless, one can get a sense of the 'lay of the land' by considering only a handful of our broadest (i.e., covering the greatest range of phenomena) and most successful theories. The set of theories I have in mind is depicted in Figure 16.1. There, each theory is shown as a region corresponding to the range of phenomena for which it is applicable. By 'applicable' I just mean that the theory does a good job of describing, explaining, and predicting the phenomena. It may be the case that there is really only one true theory of physics and the rest are ontologically misguided approximations. I won't worry about that question here, and the reader is urged not to take the drawing seriously as a claim about the structure of the world. Instead, you should think of the map as showing which theory a physicist is likely to reach for if asked to calculate a prediction or offer an explanation of some phenomenon.

To make this map, I have compressed all the characteristics of physical events into just two dimensions. Along the horizontal scale, the characteristic length of a physical phenomenon or interaction increases to the right. What I mean by a 'characteristic length' is something like the usual size of the objects involved (e.g., the radius of a planet or of an atom), or the typical distance over which a certain kind of force has an appreciable effect (e.g., the distance over which the force binding the nuclei of atoms together drops off to a negligible fraction). So theories to the left on my map are relevant for the physics of very tiny things interacting over very tiny distances, while theories on the right describe the interactions of very big things interacting over vast distances. The vertical axis on the map is intended to indicate energy. The lower you go, the less energy is involved in an interaction; the higher you go, the faster things move, the more strongly they interact, the more mass they have, and thus the more energy is involved. Note that one of the theories on my map, namely Quantum Gravity (QG), does not yet exist. That is, we know that there must be a set of laws different from either general relativity or quantum mechanics that covers phenomena of very high energy and small scale. But we don't know what those laws look like. You should also be aware that this map is at best a heuristic. It is quite artificial to arrange theories on the basis of only two quantities,

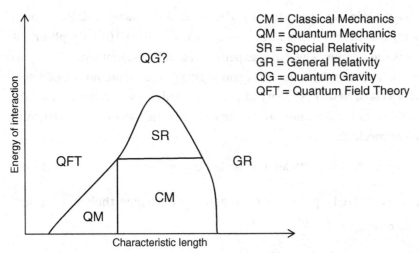

Figure 16.1   A map showing the domain of phenomena (in terms of length and energy) to which each major physical theory can be successfully applied

energy and length – they overlap in all kinds of complicated ways. For starters, the boundaries between regions are hazy. The non-relativistic theory of Quantum Mechanics (QM),[1] for instance, is needed to explain the properties of chemical reactions at room temperature, and Quantum Field Theory (QFT) is also used to describe interesting material properties at everyday energies. Even with respect to length and energy there are no sharp boundaries between the ranges of phenomena covered by different theories. It isn't as if there's an official length beyond which physicists are required to carry out their computations with General Relativity (GR) instead of Classical Mechanics (CM). Furthermore, multiple theories are often required to account for a given phenomenon. Understanding the size and stability of neutron stars, for example, requires both GR and quantum theory. So you should treat the patches I've drawn as merely suggestive of how different theories relate. Finally, I have left out major branches of physics altogether. For instance, you won't find thermodynamics on my map. This isn't because the theory is outdated or not useful. It's just because it is generally not relevant to fine-tuning arguments and because it's hard to fit it into the energy–length landscape I've chosen to draw.

---

[1] I use QM to refer to the non-relativistic theory of the quantum mechanics. Relativistic effects and systems with infinite degrees of freedom are treated by QFT.

We'll begin our brief tour at scales of around 1 meter (think of a hula-hoop or a meter-stick) and at low speeds (less than 10,000 mph or so). This is the regime of everyday experience, and for phenomena of this sort Newton's mechanics is king. As you perhaps recall from an introductory physics class, Newton's theory can be presented as a set of three laws of the sort relevant to fine-tuning arguments. Here is one way of stating Newton's laws of mechanics:

(1) Free particles (particles not under the influence of a force) do not accelerate.
(2) Objects acted upon by a force accelerate (change their momentum) according to: $\vec{F} = \dfrac{d\vec{p}}{dt}$.
(3) For an isolated system of bodies, a force exerted by one body on another results in an equal and opposite force on the first: $\vec{F_1} = -\vec{F_2}$.

Of course, Newton's laws of mechanics by themselves don't tell us much. We need to know what sort of forces exist between bodies if we are to predict how objects will move. Newton's most important contribution in this category is his famous law of universal gravitation:

$$\vec{F_G} = -G\frac{m_1 m_2}{r^2}\hat{r} \tag{1}$$

In words, Equation (1) says that two massive bodies attract each other with a force that is proportional to their masses and inversely proportional to the square of the distance between them. Put another way, more massive objects pull on one another more strongly, and the force of gravity gets weak quickly as objects move apart.

Newton's laws revolutionized the nascent science of mechanics and allowed for the precise prediction of an enormous range of phenomena, most notably the motions of the planets. Nearly a century after the publication of Newton's *Principia*,[2] another set of physical phenomena were similarly brought under the umbrella of a precise mathematical formulation, namely the phenomena of electricity and magnetism. While Newton's laws were applicable to the motions of charged objects, they had nothing to say

[2] Newton's *Philosophiæ naturalis principia mathematica* (The mathematical principles of natural philosophy) was first published in Latin in 1687. For a modern translation of the text, see Newton 1995.

about the forces between them or of the subtle interplay between electric and magnetic forces. One such force law added to the canon of physics by Coulomb in the 1780s concerns the force exerted by one charged particle on another:

$$\vec{F}_C = k \frac{q_1 q_2}{r^2} \hat{r}. \tag{2}$$

The symbols q_1 and q_2 stand for the charges of particles 1 and 2 respectively, while r is the distance between them. The symbol k stands for Coulomb's constant. The first thing to notice is that Coulomb's law is nearly identical in form to Newton's gravitational law, though charges can repel as well as attract depending on the sign of their charge (like charges repel, opposite charges attract).

These two force laws suggest a natural dimensionless constant that indicates the relative strengths of the two forces acting on matter. As it turns out, all electric charge in nature comes in integer multiples of a fundamental unit of charge, call it e. Both of the particles that make up a hydrogen atom – the proton and the electron – have a charge of exactly e: the positively charged proton has an electric charge of $+e$ while the electron has a charge of $-e$. These two particles attract each other with both an electric and a gravitational force. Using (1) and (2) above, we can compute the ratio of the electric to the gravitational force between a proton and an electron as follows (at any distance the ratio will be the same):

$$f = \frac{|\vec{F}_C|}{|\vec{F}_G|} = \frac{ke^2}{Gm_p m_e} \approx 2 \times 10^{39}. \tag{3}$$

Notice that there are no units on f. It is a pure number that does not depend in any way on our choice of scale or method of measuring electric and gravitational force. Notice also that f is enormous. The force of electricity between two typical charges is many, many times greater than the force of gravity (by a factor of 1 followed by thirty-nine zeros!). It is so great in fact that we can safely assume that gravity is absolutely irrelevant to understanding the structure of the atom. Furthermore, this dimensionless constant f tells us that the universe would be vastly different if matter wasn't on average electrically neutral. Gravitational interactions would be completely swamped by electric forces, and the large-scale structure of the universe would depend mostly on the forces between charges.

So after taking a quick look at some of the laws of physics that apply at the scales of length and speed we're used to, we were able to identify a dimensionless constant that told us something about the way these laws shape the universe. Let's now turn our attention to the world of little things. In the next section, we'll move to the left on my map and consider some of the laws that govern physical phenomena at the scale of atoms and below.

16.3 A survey of natural law: particle physics and the very small

As we adjust our focus downwards to smaller and smaller scales, we move into the world of the quantum. The theories describing physical phenomena at the smallest scales are very different from Newtonian mechanics. Unfortunately, I only have space enough to tell you what theories apply at these small scales, and to suggest a few important numbers that determine the character of the phenomena they describe. For a more complete but accessible overview of quantum mechanics I suggest chapters 43 and 44 of *Fundamentals of Physics*.[3] For a largely non-technical sketch of the remaining theories, you might look at Richard Feynman's *QED*.[4] For a more rigorous survey of every theory discussed in this chapter, you can turn to the excellent book by Roger Penrose, *The Road to Reality*.[5]

First, let me give a sense of the scale at which Newtonian physics ceases to yield good predictions. The human body, which taken as a whole is well described by Newtonian mechanics, is made up of around 100 trillion (10^{14}) cells, each of which is around 10 millionths of a meter (10^{-5} m) in diameter. Within these cells, most functions are carried out or regulated by very large organic molecules called proteins. A large protein, like RNA polymerase, is around 10 billionths of a meter (10^{-8} m) across. Even at this tiny scale, matter can still be reasonably well modeled using Newton's mechanics and Maxwell's electrodynamics.

It is only when we reach the scale of the atom at around 1 Angstrom (10^{-10} m) that the phenomena diverge wildly from what Newtonian mechanics predicts. At this length-scale, QM and the related theory of quantum electrodynamics (QED) prevail. The latter theory describes how

---

[3] Halliday *et al.* 1993.   [4] Feynman 1985.   [5] Penrose 2004.

the tiny charged particles making up an atom interact with one another and with the electromagnetic fields in which they find themselves.[6] An important dimensionless constant arises from considerations of the interaction of fundamental particles with the electromagnetic field. This constant, called the 'fine-structure constant' and usually denoted by α, appears in QED as a measure of how strongly electromagnetic fields interact with the massive charges that produce them. You can think of it like the volume knob on your stereo. This knob determines how strongly variations in voltage from your radio are coupled to motions of the speaker. The higher the volume, the more energetically your speakers respond to the same signal from the radio. The fine-structure constant determines how energetically electrons and other charged particles respond to electromagnetic fields. The bigger α is, the more energetically charged particles respond to the field.

If the value of α were very different from what it is now (while keeping everything else fixed as far as possible), the world we know would probably be impossible. For instance, weakening α just a little would mean that light in the visible wavelength would not be energetic enough to excite the electronic transitions in chlorophyll which are essential to photosynthesis – there could be no plants. Similarly, making α too strong would mean that even modest radiation (e.g., sunlight) would be sufficient to blast apart big molecules. It is hard to see how life could exist in that case.

The fine-structure constant α is what's known as a 'coupling constant'. It indicates how strongly fields are coupled with charged particles.[7] At the scale of the nucleus and below there are two more sorts of physical fields which carry force: the strong and weak fields. Each of these has a coupling constant of its own (denoted α_S and α_W respectively) which determine quite a lot of the character of matter at small scales and high energies. Like α, these coupling constants cannot be predicted by any known theory and amount to 'free parameters', numbers we have to put into the theory as basic facts. These are just the sorts of things that folks worry about being 'fine-tuned' for life.

---

[6] As far as QED is concerned, the nucleus is a single charged particle. To describe the interactions amongst the particles making up a nucleus, one must turn to quantum chromodynamics (QCD). But nuclear interactions are largely irrelevant to chemistry.

[7] More accurately, it indicates how strongly fields couple to one another. In QED, particles are themselves just modes of a field that carry mass.

16.4 A survey of natural law: cosmology and the very large

We've considered some of the most important constants characterizing our physical theories of the very small. Let's turn our attention now to the very big. How big? Well, we know that there is structure in the universe on a vast scale. For starters, there are planets. These rocky or gaseous objects range in size from around 10^6 to 10^8 meters in diameter (from a few thousand to hundreds of thousands of miles in diameter). Stars are even bigger (around 10^9 meters), and solar systems made up of planets and stars span some 10^{13} meters. Solar systems populate galaxies, enormous conglomerations of stars about 10^{21} meters across. Galaxies in turn cluster together into groups of about 10^{22} meters across, which in turn assemble into titanic superclusters 10^{24} meters across. To put this in perspective, the size of our solar system relative to a supercluster is like the size of a hydrogen atom compared to you.

On the scale of planets and up, gravity is the dominant force[8] and Einstein's theory of General Relativity (GR) is the relevant physical law for explaining motion. It is also the foundation of physical cosmology, the scientific theory of the structure and evolution of the universe. Since cosmology is often invoked in fine-tuning arguments, we will need at least a cursory grasp of Einstein's theory to follow the claims made in these arguments. In the simplest description, GR is a law relating the 'metric' of spacetime to the distribution of matter and energy. A metric is a measure of the distance between two points in space, whether that space is physical (like the three-dimensional space we move around in) or abstract (like the 'space' of possible configurations of particles). You probably learned one such metric in school: the Euclidean measure of distance. On a plane – a flat two-dimensional surface like the top of a table – the Euclidean metric is given by:

$$\Delta s = \sqrt{\Delta x^2 + \Delta y^2}. \tag{4}$$

Here, Δs is the distance between the two points of interest, while Δx and Δy is the difference between their x-coordinates and y-coordinates respectively.

---

[8] This is only because overall large bodies tend to be electrically neutral. If they weren't, electric and magnetic forces would dominate.

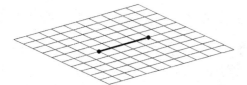

Figure 16.2   A plane surface on which are indicated two points and the shortest path between them

Technically, Equation (4) is not a 'differential metric' of the sort we'll need to do GR, but we can make it one with a little mathematical finagling. If we replace the finite differences in coordinates indicated by the 'Δ' with infinitesimal differences of the sort you see in calculus, we can write:

$$ds^2 = dx^2 + dy^2. \tag{5}$$

What Equation (5) tells us is that the differential distance between nearby points on our surface is given by the old Euclidean relation. To get actual distances between any two points, we would have to integrate Equation (5) – to add up all the infinitesimal distances – from one point to the other along some path. We call the length of the shortest such path the 'distance' between the points. If all this math is opaque to you, just think of metrics like the one shown in Equation (5) as a little computer program into which you can input pairs of points and a path between them and it will output the distance between them along that path. Figure 16.2 depicts the sort of plane surface or tabletop geometry we've been talking about and indicates a path between two points on this surface. In this case, if we integrate the metric (5) along this 'straight path' the answer we get is just that given by Equation (4).

So why bother with the metric at all? Why can't we just use our old Euclidean formula, Equation (4)? The answer is that we are interested in other sorts of geometries. For instance, we happen to live on the surface of a globe. If you have to stick to the surface, then the shortest path between two points – say Pittsburgh and Doha – does not have a length given by Equation (4). The geometry of the surface of a sphere is just not like the geometry of a plane. For one thing, the interior angles of a triangle do *not* sum to 180°. So how do we compute the relevant distances? How do we determine the lengths of arbitrary paths on the surface and the angles between intersecting lines? We write down an appropriate metric. In this

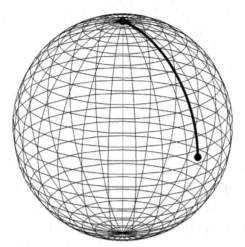

Figure 16.3   A spherical surface on which are shown two points and the shortest path between them (i.e., the 'straight line' on a sphere)

case, it will be easier to use so-called 'spherical coordinates'. Instead of specifying a position with three numbers, x, y, and z, that stand for positions
along three mutually perpendicular axes, spherical coordinates indicate
position by using two angles, θ and φ, to pick out a direction and a length r
indicating how far out from the origin a point is in that direction. In these
coordinates, the metric for a sphere of radius R is given by:

$$ds^2 = R^2 \left(d\theta^2 + \sin^2\theta d\varphi^2 \right). \tag{6}$$

Figure 16.3 shows the surface of a sphere and a sample path between two
points. Note that in three dimensions, this path appears curved – it is not
the shortest path. But if we have to stick to the surface of the sphere, just
as you and I have to stick to the surface of the Earth, then it is the shortest
path. That 'curve' is a straight line on the surface of a sphere.

A metric can tell us about geometries which are not flat, which vary
from place to place, or for which the shortest path is *not* equivalent to the
Euclidean straight line. They can also help us get a handle on geometries
that change with time. Suppose now that, instead of thinking of a flat plane
surface, we consider the sort of three-dimensional world in which Euclid
thought we lived. That is, let's add a dimension – represented by a variable
z – to our plane surface so that the metric becomes:

$$ds^2 = dx^2 + dy^2 + dz^2. \tag{7}$$

This metric implies that the shortest distance between any two points is the sort of straight line you are familiar with. In fact, the distance between any two points, after we integrate the metric in Equation (7), is just the Euclidean formula, $\Delta s = \sqrt{\Delta x^2 + \Delta y^2 + \Delta z^2}$. But now suppose that we introduce a time-dependence into our metric:

$$ds^2 = a^2(t)(dx^2 + dy^2 + dz^2). \tag{8}$$

In our new metric, the function $a(t)$ tells us how distances between the same two points vary with time. This is often called 'raisin-bread metric' since if we let $a(t) = kt$ the metric describes the geometry inside a loaf of baking raisin-bread. Imagine we put our raw dough in the pan at time t_0. As the dough bakes, the leavening agent emits carbon dioxide gas more or less evenly throughout the dough, causing it to expand in all directions. Let's consider how the distance between any two raisins changes with time. At the beginning, suppose the raisins are separated by a distance $d_0 = a(t_0)\sqrt{\Delta x^2 + \Delta y^2 + \Delta z^2}$. At some later time t_1 the raisins have moved apart and are separated by $d_1 = a(t_1)\sqrt{\Delta x^2 + \Delta y^2 + \Delta z^2}$. That is, *for any two raisins the distance between them at t_1 is* $\dfrac{a(t_1)}{a(t_0)}$ *times bigger*. According to our metric, every raisin is moving away from every other raisin at a speed proportional to the distance between them. Thus, the further away one raisin is from another, the faster they are moving apart. Specifically, the speed v at which the raisins move apart is given by:

$$v = hd. \tag{9}$$

In Equation (9), d is the distance between the raisins and $h = \dfrac{k}{a(t)}$, where k is the constant in the metric function $a(t)$.

As it so happens, if you look out into the universe from Earth and measure the speeds of all the galaxies you can see relative to our own, you will find, as Hubble did, that they are all moving away from us and doing so at a speed proportional to the distance between them:

$$v = H_0 d. \tag{10}$$

In this case, H_0 is known as 'Hubble's constant' and, like the parameter h in my raisin-bread example, is not really a constant – it is a function of time. However, it changes so slowly relative to what we can observe, that for all observational purposes you can think of it as a constant. Now, the best-known model of the universe we can construct with GR that accounts for this Hubble relation tells us that the universe essentially has a raisin-bread metric:

$$ds^2 = -dt^2 + a^2(t)(dx^2 + dy^2 + dz^2). \tag{11}$$

This is the Friedman–Robertson–Walker metric. You'll note that unlike the metrics I introduced for thinking about space this one has a dt in it. In GR, space and time are treated as geometric coordinates on an equal footing because it turns out that you can't talk about an objective distance in space that is the same for all observers – you have to consider distances that are a combination of space and time intervals. To get a hint of why this is the case, consider two of the most famous phenomena predicted (correctly) by relativity theory. First, objects moving very fast are contracted in length. That is, a meter-stick flying parallel to the ground at near the speed of light is shorter than 1 meter in length according to an Earth-bound observer. Second, an observer riding on the flying meter-stick will experience shorter time intervals between events than an observer on Earth. These effects combine in such a way across different frames of reference that, though we cannot agree on distances or time intervals, we can agree on the value of a particular combination of time and space intervals.

I do not have space to try and clarify this profound shift in characterizing the world, but can only encourage you to seek out additional sources (such as Einstein's popular book on relativity[9]). For now, it is enough for you to understand that there is a crucial dimensionless constant in GR that determines whether the universe has a metric that looks like Equation (11) or one that looks more like that of a sphere. This constant – called the 'expansion coefficient' and represented by Ω_0 – is the ratio of two densities: the critical density of matter in the universe ρ_{crit} for which Equation (11) is the correct description of the universe and the actual density of matter in the universe ρ_0:

$$\Omega_0 = \frac{\rho_0}{\rho_{crit}}. \tag{12}$$

[9] Einstein 1961.

The expansion coefficient Ω_0 is the last dimensionless constant we'll consider, the principal constant dictating the structure of the universe at very large scales.

This concludes our tour of physical law. We looked briefly at the particle theories thought to govern the world at its smallest scales and emphasized some of the important dimensionless constants that play a role there. Likewise, we canvassed the laws that determine the structure of space and time itself at very large scales and considered what dimensionless parameters must be inserted into the theory to give descriptions matching the world we have. We'll meet these parameters again in the next two chapters.

17 Fine tuning I: positive arguments

17.1 A cosmic conspiracy?

In 1979, a curious paper appeared in the pages of the eminent scientific journal *Nature*.[1] The paper was unusual because it did not report on novel experimental findings. It did not offer a new theory or explanation of known phenomena. It did not even survey the current status of a developing branch of science as other 'review articles' in the journal typically do. Instead, it argued for a long list of coincidences of two kinds. First, there are the connections between different branches of physical science. As the paper's authors, B. J. Carr and M. J. Rees explain, "[t]he structure of the physical world is manifested on many different scales, ranging from the Universe on the largest scale, down through galaxies, stars and planets, to living creatures, cells and atoms ... Each level of structure requires for its description and explanation a different branch of physical theory, so it is not always appreciated how intimately they are related."[2] They set out in part to exhibit this intimate connection by demonstrating how most of these disparate 'natural scales' described by different theories are all determined by a shared handful of dimensionless parameters. In fact, most of their examples involve just three parameters: the electromagnetic fine-structure constant (α), the gravitational fine-structure constant (α_G),[3] and the electron-to-proton mass ratio (m_e/m_p). So, for instance, the authors provide a simple physical argument for the fact that most stars have a mass in a range between 0.1 and 10 times that of our Sun. This range of possibility is itself dictated by the parameter α_G. Of course, that parameter might have had a different value, at least as a logical possibility. In that case, the range

---

[1] Carr and Rees 1979.   [2] Carr and Rees 1979, 605.
[3] I didn't define this parameter in the last chapter. It is given by $\alpha_G = Gm_p^2/\hbar c$.

of masses computed would be different. So whatever value α_G has, physical processes of the sort operating in our universe guarantee that stars form only in a particular range, but what that range is could be different. Similarly, the authors argue that the size of a typical atom is dictated by the value of α – the bigger α is, the smaller a typical atom is.

But many of the coincidences discussed by Carr and Rees are of a different sort – they are coincidences that link the values of more than one dimensionless parameter in relations that must hold if the universe is to harbor life as we know it. These relations involve features of the physical world that are not consequences of the value of any single parameter, but instead are consequences of the way in which multiple parameters balance one another. Aside from their implications for life, what's notable about these relations is their fragility. Here's what I mean. Physical arguments like those given by Carr and Rees tell us why the world is as we find it, given the values that the parameters actually have. These same arguments tell us that unless the value of each parameter falls within a surprisingly narrow range, the world would be a very different place. In particular, it seems that varying one parameter just a little while holding the others fixed would make the universe inhospitable to life.

Let's consider one such chain of reasoning using the examples Carr and Rees provide. While it is true that we can imagine a universe in which α_G is much larger, it is not the case that this universe would look just like the one we have except with lighter stars. Rather, the lighter stars would be purchased at the price of habitable planets! This is because, in order for planets to form with thick atmospheres dominated by gases other than hydrogen, the ratio of parameters α/α_G must fall within a narrow range. This means that, in order to preserve features of the universe that (we think) are essential for life, we would have to increase α while increasing α_G. But this, in turn, would have other unpleasant consequences. In this case, it would mean dramatically decreasing the sizes of atoms and altering chemistry in rather profound ways that would still make life as we know it implausible if not impossible. Or consider instead the relationship between the parameter Ω_0 that I described in Chapter 16 and stellar lifetimes. The parameter Ω_0, called the expansion coefficient, determines amongst other things the shape and age of the universe. The lifetime of a main-sequence star like our Sun is, Carr and Rees argue, determined by α_G and the mass of the proton. If Ω_0 were changed to be a little above or below 1 while leaving the other parameters fixed, then the universe would either have collapsed before any stars could mature, or would have expanded so quickly that the primordial gases could never have pulled together into stars. As Carr and Rees sum up the situation, "[S]everal aspects of our Universe – some of which seem to be prerequisites for the evolution of any form of life – depend rather delicately on apparent 'coincidences' among the physical constants."[4] This delicate dependence is now universally referred to as the *fine tuning* of the parameters of physics.

17.2 Anthropic principles, physical coincidences, and design arguments

People have had dramatically different reactions to the kinds of fine tuning fingered by Carr and Rees. On one end of the spectrum, there are those who react with disbelief that anyone is puzzled by something so obvious. Of course the parameters have the values required for life. Otherwise, we wouldn't be here to marvel at those values! To adapt a quip from Abraham Lincoln, we might capture this reaction with an analogous observation: it isn't very surprising that my legs are just long enough to reach from my body to the floor. What the coincidences noted by Carr and Rees and many others after them indicate is only a sort of selection effect, not a profound fact that cries out for explanation. Obviously, we as human beings can only make observations that are compatible with the fact that we are human observers. In the terminology of John Barrow and Frank Tipler,[5] here's one way to capture this skeptical response in the form of a principle:

> **Weak Anthropic Principle (WAP)**: The observed values of all physical and cosmological quantities are not equally probable but they take on values restricted by the requirement that there exist sites where carbon-based life can evolve and by the requirement that the universe be old enough for it to have already done so.

WAP amounts to the assertion that if observation O is found to be the case, then we know that any theory T that implies that O is false (or really unlikely) cannot be true (or probably isn't). Of course, calling this a principle may be a bit of a stretch. It is difficult to see how WAP is distinct

---

[4] Carr and Rees 1979, 605.   [5] Barrow and Tipler 1986.

from inductive reasoning in general – science consists in large measure of reasoning from effects to possible causes, excluding some possible causes on the basis of known effects. WAP just emphasizes one particular empirical fact, namely that all empirical facts on which science is based were acquired by human observers. But whatever its merits and status as a principle, WAP can be read as a sort of attempt to explain away the cosmic coincidences by pointing out that we simply wouldn't be here to discover the coincidences amongst the parameters characterizing the laws of physics unless those parameters fall within the range that can support life. We will dwell more on this line of reasoning – often called 'anthropic' reasoning – in the next chapter.

But there is another line of anthropic reasoning, one that sees the cosmic coincidences described by Carr and Rees among others as implying a much stronger claim. There are both physicists and philosophers who subscribe to a stronger version of WAP along these lines and embrace the following principle:[6]

Strong Anthropic Principle (SAP): The universe must have those properties which allow life to develop within it at some stage in its history.

One way to motivate this view is to accept a more fundamental principle articulated by Paul Davies.[7] Davies points to what he calls a "mediocratic progression" that begins with Copernicus teaching us that Earth is an unexceptional example of a planet orbiting the Sun. Equipped with better and better telescopes, later astronomers learned that our Sun is but a mediocre instance – a typical example – of a star amongst an enormous number of such stars. Likewise, our galaxy is but one typical example in a universe filled with galaxies. Extrapolating by a weak sort of induction, one might be inclined to embrace the further claim that there cannot be anything atypical or special about our circumstance as observers; we must be typical observers in a typical universe. In other words, most possible universes look like ours. This would entail SAP (or at least a probabilistic version of it).

---

[6] Barrow and Tipler present one more even stronger (and stranger) principle, which Martin Gardner charmingly referred to as the Completely Ridiculous Anthropic Principle, or 'CRAP' (Gardner 1986).

[7] Davies 2007, 129–31.

Whatever reason leads to its adoption, SAP is not the truism that WAP is. It asserts a much stronger claim about what sorts of universes are possible irrespective of the empirical facts. Certainly if SAP were true, it would explain the coincidences we see. But SAP raises more questions than it answers. What compels the universe to turn out this way? Why would the conditions for life determine what physics is possible? For many, the answer to these questions is intuitively obvious, driven by an overwhelming intuition of conspiracy of the sort we discussed back in Chapter 1. Reflecting on a coincidence in physical constants that permits the production of roughly equal amounts of carbon and oxygen in stars, the astronomer Fred Hoyle exclaimed: "A common sense interpretation of the facts suggests that a superintellect has monkeyed with physics, as well as with chemistry and biology, and that there are no blind forces worth speaking about in nature."[8] In other words, SAP and the various coincidences it explains can all be accounted for as products of design. Of course, there are other possibilities, and we'll see some of them offered as objections to design arguments based on fine tuning. Other possibilities include the existence of many universes (of which ours is just one) and the logical necessity of the laws of physics (and thus the trivial truth of SAP). This is a live debate with few points of agreement, but one widely endorsed claim is that fine tuning cries out for explanation. Once we accept this, then we must concede that the existence of God or gods might be the best (or only) explanation.

17.3 The nature of physical law

In the previous chapter, I provided a short overview of some of the laws of physics as they are currently understood. In doing so, I spent a great deal of time talking about the content of physical law. But I did not say much about what sort of a thing a law of physics or a 'law of nature' is in the first place. In one sense, the answer is obvious – a law is just a true generalization like all X are Y or all massive bodies attract one another according to the inverse square of their separation.[9] It is a rule one can write down as a sentence or equation. But is there anything more to being a law of nature?

---

[8] Hoyle 1982.
[9] Whether we should reserve the term 'law' to refer only to universal generalizations is controversial, but this is a merely semantic debate.

Broadly speaking, there are two competing answers to this question.[10] One answer is that a law is just a summary of facts. It is a compact way of stating what has occurred in the past and of what will occur in the future. In other words, laws are descriptions of the regularities that are actually to be found in nature.[11] I'll call this the *Regularity view* of physical law. On this view, the universe is a collection of events. There are many true propositions we can assert about these events, many regularities that hold amongst them. But which ones we elevate to the status of a law may be partly accidental (depending on what we notice or have the ability to detect) and partly conventional (in that we could have chosen to group what we do know in different ways). Laws in this view are objectively true statements about the world of events, but there may not be a unique 'best' set of laws. There are many subtle variations and elaborations of Regularity positions. What's important to note for our purposes is that in all of the Regularity accounts, the facts of the world determine the laws and not vice versa.

The Regularity view has fallen into disfavor. Its critics cite a number of problems for the position. Chief among them is the claim that a Regularity view cannot account for the intuitive difference between a generalization that is 'accidentally' true of the phenomena and one that is genuinely a law. Let me recount a standard example originally introduced by Bas van Fraassen.[12] On the one hand, it is probably true that there do not exist any gold spheres with a radius of 1 km. It doesn't seem to be impossible. There doesn't seem to be anything preventing one from going about the universe collecting gold in sufficient quantity to make such a sphere. But nonetheless, there is reason to suppose no such spheres actually exist. On the other hand, that there are no spheres of uranium 235 (the fissile isotope of uranium) with a radius greater than 1 km seems to be necessarily true. Anyone who tried to make one would be very sorry very fast (the amount of uranium 235 brought together in an atomic bomb is miniscule by comparison). Both generalizations about the non-occurrence of certain objects are true. On the Regularity view, they have

---

[10] See Psillos 2002 and Swartz 2009 for helpful overviews of the debate about laws of nature.

[11] On this view, we can still be wrong about what the laws are. They are not merely reports about the observations we have made or expect to make. Rather, they are universal generalizations that may fail to be true of the world – we can incorrectly guess or infer laws from our finite data.

[12] van Fraassen 1989.

equal status. But for many, it is intuitively the case that there really is a law about uranium spheres, but no such law in the case of gold – it is just an accidental fact that our universe contains no giant golden spheres.

An alternative (and, at the moment, more popular) view of physical law is one I'll call the *Necessitarian view*. This view asserts that laws in some sense exist prior to the events of the world and determine, at least partly, the way in which these events unfold. Under the Necessitarian view of laws, there really is a unique set of laws we can write down, each of which actually corresponds to a genuine thing out there 'in the world' determining or governing the way the world unfolds. In addition, events are supposed to conform to laws necessarily. The Necessitarian view emphasizes this relation of necessitation, whatever exactly that may mean. It is this emphasis on necessity that allows this view of laws to distinguish clearly between universal claims that are accidentally true and those that are lawlike. Accidental generalizations may be true but are not necessitated by any laws – they could have been different in a world with precisely the same laws of nature. Lawlike generalizations on the other hand are necessarily true because they are the consequence of some law. They must be true in all worlds governed by the same laws. On the other hand, the laws themselves are generally thought to be contingent – one can imagine possible worlds governed by different laws.

Clearly, if we are to talk about rigging the laws of nature or fine tuning physics for the purpose of producing life, we must have in mind a Necessitarian view of physical laws. It makes no sense to speak of adjusting the constants in the laws to produce particular effects unless those laws exist independent of the effects they determine. So the proponent of fine tuning rules out certain interpretations of what a physical law is from the outset. This is a potential cost of accepting such an argument that we will have to be kept in mind. Additionally, different positive arguments from fine tuning will implicitly adopt different sets of supplementary metaphysical ideas concerning the nature of laws. Watch out for these – any neutral evaluation of a fine-tuning argument must bring them under scrutiny.

17.4 Davies' fine-tuning argument

The physicist Paul Davies makes no bones about assuming a Necessitarian view of laws, declaring that "[t]he laws of physics ... *really exist* in the world out there, and the job of the scientist is to uncover them, not invent them."[13] This stance concerning what it is to be a law of nature plays a central role in his design argument from fine tuning, the first such argument we'll consider.[14] Davies' argument has two parts. The first, represented as a diagram in Figure 17.1, is an argument to the conclusion that the laws of physics must have a cause, something by virtue of which they are as they are. This part of the argument rests on some substantial metaphysical claims with a decidedly medieval flavor. In fact, the important premises concerning contingency are shared with Thomas Aquinas, and appeared in Aquinas' cosmological argument (one of the Five Ways mentioned in Chapter 4). In particular, Davies implicitly assumes that every state of affairs – in this case the laws of physics having the form they do – must be either necessary or contingent. A necessary state of affairs is one that could not have been otherwise, while a contingent state of affairs could have been otherwise (or simply not have existed at all). According to Davies, we know that the laws of physics are contingent because other laws are possible, and we know other laws are possible because we can consistently imagine them. Of course, this last step means embracing the very old premise that conceivability entails possibility. If we borrow one more premise from Aquinas, namely that every contingent state of affairs must have a cause, the contingency of the laws of physics implies the existence of a cause of the laws of nature. By 'cause' I do not necessarily mean an event that preceded the laws in time, but rather another state of affairs on account of which the particular laws in question hold for our universe. Without digging too deeply into the metaphysics, Davies thinks that the contingent nature of the laws implies a cause, and it is in this cause that we shall find God.

The second half of Davies' argument is where fine tuning comes into the picture. What he gives us resembles the old argument from order. In that old argument, an exhaustive list of possible causes for a property is presented, and then all but design are eliminated. For the Stoics, the property in question was the regularity of celestial motion and there were only two possible causes or explanations: chance and design. In this case, we aren't considering a property of things in the universe. Rather, the object of explanation, the thing for which we seek a cause, is a property of the

---

[13] Davies 2003, 149; emphasis in original.
[14] The argument presented appears in Davies 1992 and 2003.

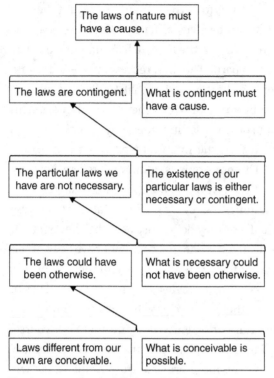

Figure 17.1   Davies' fine-tuning argument

very form of the laws of physics that in turn explain everything within the universe. The property in question is fine tuning. Why are the dimensionless parameters that characterize our laws so precisely configured for life, beauty, complexity, mind, and other phenomena of inherent value to arise? Again, there is a short list of possibilities. First, there is design. Since, says Davies, things of inherent value such as life are precisely what a deity would aim at creating, the hypothesis of design does a good job of explaining why the laws appear to be rigged for the production of value.

What is the alternative? What is the analog of the chance hypothesis? In this case, the alternative causal account invokes multiple universes. Roughly, the idea is that there exist a great many (perhaps infinitely many) separate universes. The ensemble of all universes is called the 'multiverse'. Each universe in the multiverse comes into existence with a possibly distinct set of values for its dimensionless parameters, and therefore a possibly distinct set of laws of physics. In the simplest scenario, these values are randomly assigned. In that case, there is a high probability that one of the universes has laws that allow for life like us to exist. It should therefore come as no surprise that we are asking these questions from within such a universe. So the multiverse hypothesis explains the apparent fine tuning in a two-step fashion. First, it provides a physical model in which our universe is in some sense unexceptional – in an infinite ensemble of universes there are likely quite a few that support life. Second, we appeal to something like WAP to explain why we happen to find ourselves in an uncommon kind of universe.

As Davies notes, there are many variations on the multiverse theme. I'll describe two proposals that together give a sense for the full range of possibilities. We'll begin with bubble universes. According to the dominant cosmological model, our universe began with a bang – the so-called 'Big Bang'. More accurately, at early times, space was significantly smaller or more compressed than it is now (distances between any two points were smaller), and all matter was hot and dense. For the first tiny interval of time, space expanded exponentially in a process with the descriptive if understated name of 'inflation'. Now, it is possible – though certainly not required by the theory of inflation – that in this process of expansion, different regions of the universe slowed their expansion at different times. The result is a collection of regions, each of which is more or less stable like the observable universe in which we find ourselves. However, each of these regions is separated by a vast expanse of space that continues to expand rapidly. These distinct regions are so far separated and the space between them expanding at such a rate, that not even light can reach between them. In other words, they are completely causally disconnected islands, all occupying the same arena of spacetime. Now, to speculate even further, we can imagine that in each of these condensed regions or island universes, the dimensionless parameters of physics take on different values (mechanisms for such a process have been proposed in the physics literature). There are, of course, problems with using this model to explain the fine-tuning phenomenon. For one, such a model requires us to posit the existence of bubble universes which we cannot detect, even in principle. This, says Davies, is to ignore Occam's razor, the dictum that we should favor the theory that posits the fewest entities (the 'simplest' hypothesis). This multiverse hypothesis requires that we posit the existence of infinitely many universes to explain what the design hypothesis accomplishes with a single entity.

So let's consider an alternative multiverse model. The physicist Lee Smolin has proposed a model of 'cosmological natural selection'.[15] In this model, a diverse population of universes is created as universes bud off one another through black holes. I won't attempt to sketch the speculative physics that makes this remotely plausible, but this idea is distinct from the bubble universes in that the proposed processes result in completely separated spacetimes (disjoint universes) rather than bubble-like regions sharing the same spacetime. As one universe buds off another, it acquires laws of physics that vary randomly with respect to its parent. Some laws of physics will result in the production of more black holes than others, and so some universes will produce more 'offspring' universes than others. Thus, in the overall ensemble, there will be a sort of selection process, supposedly akin to biological selection. Universes good at making black holes will dominate. Universes that make black holes are also universes with laws amenable to life. Thus, since black-hole-producing universes are highly probable, so too are universes with laws just right for life. Again, there are a variety of problems with this model, but I'll focus on the one Davies emphasizes. In Smolin's model – in fact, in any multiverse model – we have supposed a set of meta-laws that determine how the laws of each baby universe are determined. But this means that we are left with a puzzle very much like the one we started out trying to solve. Why are the meta-laws precisely as they need to be to produce universes that support life?

A number of rebuttals suggest themselves. For instance, it may be the case that a wide range of meta-laws would nonetheless result in a process of cosmological natural selection favoring life-producing laws of physics. This depends on the details of Smolin's model, and so I won't pursue it. Alternatively, one might object that the hypothesis of a designer suffers the same sort of problem: why would God have the sort of goals that would lead him to create a universe with laws that allow for the possibility of life? Setting these possible rebuttals aside, the upshot for Davies is that, of the two alternative explanations, we have reason to reject the multiverse and therefore reason to embrace design.[16]

[15] Smolin 1997.

[16] In his later work, Davies favors a different hypothesis. Specifically, he argues that we should accept that the universe is 'self-explaining' rather than designed (Davies 2007).

17.5 Craig's argument from specified complexity

Whereas Paul Davies presents an argument based on medieval notions of contingency and necessity, the philosopher William Craig attempts to adapt Dembski's very modern argument from 'specified complexity' to the universe as a whole.[17] In Chapter 13 we considered the details of Dembski's argument, and I won't recount those details here. Instead, I'll present a reconstruction of the argument Craig provides, leaving the interested reader to fill in relevant details from Dembski's work where necessary.

In a nutshell, Craig's argument can be put like this: the combination of the physical laws and initial conditions that govern our universe demonstrates specified complexity. Since specified complexity is a reliable indicator of design, we can safely conclude that the laws of the universe were designed. Naturally, design on such a grand scale and outside of the normal arena of space and time in which all known events occur suggests that the designer in this case is worthy of the appellation 'God'. To make his case, Craig has to show that in fact the laws and initial conditions of this universe together meet the conditions for a thing to exhibit specified complexity. As Dembski presents it, the property of specified complexity consists of two parts: improbability and conformity to a pattern. Thus, to show that laws were designed, we have to show that they are both highly improbable and conform to a salient pattern of some sort.

Here is how Craig makes the case for specified complexity. To begin with, we note that the dimensionless parameters (which Craig calls "constants") characterizing the laws of physics along with the cosmological initial conditions (i.e. the distribution of matter and energy in the universe at the moment of the Big Bang) possess certain values. Let's call the collection of values X and the observation that the initial conditions and parameters are those in X we'll call E (for 'evidence'). X contains all of those numbers that Carr and Rees were talking about, the numbers rife with apparent coincidences. Now, no current physical theory indicates that the parameters and initial conditions must have the values in X. So, just as Davies does, Craig concludes that E is not physically necessary. According to Craig, that means it is either the result of design or 'chance'. Of course, there is something special about the values in X – they are amongst the tiny set of

[17] Craig 2003.

values for which carbon-based life is possible in the universe. We might say, then, that the occurrence of X is an instance of an especially salient pattern we'll call D*. That is, every event that exhibits the pattern D* is the occurrence of a set of physical parameters and cosmological initial conditions that are sufficient for carbon-based life to arise.

So far, we've only shown that the occurrence of the parameters and initial conditions we have satisfies a salient pattern. We haven't yet established that it exhibits specified complexity. To do so, we need to show that getting values of the parameters in X that satisfy D* is improbable.[18] Probability, though, is relative to a distribution and a space of possibilities. What is the relevant space of possibilities here? How should we distribute probability over these possibilities? Craig assumes that the relevant spaces of possibilities is a continuous range of possible values for each of the items in X (he doesn't specify this range) and that the distribution is uniform over this range. That is, every possible outcome is equally likely. Under these assumptions (given that the range of possible values we consider for the items in X is sufficiently large and the range for D* sufficiently small), then it is true that the probability of picking an X that satisfies D* is vanishingly small.

However, even under Craig's assumptions about the space of possibilities, the small probability of getting a universe with values in X could be overcome if we admit the possibility of many samples. That is, if we suppose there is a multiverse – a large or perhaps infinite set of universes each characterized by a set of values drawn at random from the space of possibilities – then it would be highly probable that at least one has the values such that X satisfies D*. But Craig argues against this possibility, ultimately concluding that, "[t]he error made by the Many-Worlds Hypothesis is that it multiplies one's probabilistic resources without warrant. If we are allowed to do that, then it seems that *anything* can be explained away."[19] With only one universe, and so only one actual sample of initial conditions, it follows that the total probability of obtaining parameter values equal to those in X

---

[18] Craig writes as though D* contains only X. In fact, he explicitly says that X and D* are identical. But this would make the probability of either strictly 0 since the probability of any single point value is 0 assuming a uniform distribution over a continuous set. To give Craig the benefit of the doubt and to remain true to Dembski's theory, I assume that D* actually contains a continuous collection of sets of fine-tuned constants and consider the relevant probability to be that of finding X within D*.

[19] Craig 2003, 173; emphasis in original.

by chance is very, very small. In fact, the probability of obtaining the values in X is small compared to any reasonable threshold for rejecting chance hypotheses, and so we may conclude that E cannot be the result of chance.

We have almost everything we need, according to Dembski's conditions, to establish specified complexity: we have an event, E, that is highly improbable and which conforms to a pattern D*. All that remains is to show that we haven't picked an ad hoc pattern, one chosen after the fact to make the event E seem important. Dembski says that we can check this by making sure that we can specify the pattern D* without reference to E. To this end, Craig claims that we have some information, Inf, on the basis of which we can characterize the range of initial conditions which would permit carbon-based life. That is, Inf lets us characterize D*. Furthermore, the information, Inf, is independent of the fact that E is true. That is, $P(E \mid H \,\&\, Inf) = P(E \mid H)$ where H is the chance hypothesis. This completes the requirements Dembski gives for specified complexity (at least in one version of Dembski's scheme). Thus, we are warranted in inferring that E is the result of design. That is, because the constants and initial conditions that characterize the physical laws governing our universe exhibit specified complexity, we may infer that these laws and thus the universe were designed by a deity.

Of course, since Craig's argument involves an application of Dembski's inference scheme, it is subject to all of the same objections and concerns. Since it involves assigning probabilities to features of the universe as a whole, it raises additional concerns as well. These will be taken up in the next chapter. For now, I want to draw out a distinction to which Craig pays little heed: the difference between initial conditions and parameters of a law.

Imagine that, instead of worrying whether the universe was designed, we consider a simpler question about golf. Suppose in particular that you learn that I have hit a hole in one on your favorite golf course. We'll denote this surprising fact (quite surprising if you've ever seen me play golf) by O. Let's consider two hypotheses that might explain O:

H_1: I was aiming at the pin and intended to hit a hole in one.

H_2: I was aiming at the green and didn't intend to land the ball in any particular place on the green.

In order to invoke the likelihood principle to favor one hypothesis over the other, we will need to compute $P(O \mid H_1)$ and $P(O \mid H_2)$. How are we to assess these probabilities? In this case, the answer is straightforward. The flight of the golf ball from the tee to the hole was governed by Newton's laws of mechanics and the law of universal gravitation. Exactly which trajectory the ball followed when I hit it was determined by the angle at which I held the face of the club relative to the ball, and the speed with which the club head was moving. Those two numbers, the angle θ and speed v at the time of impact, are examples of what are referred to as initial conditions. I can now understand my two hypotheses as providing a probability distribution over the range of possible initial conditions. H_1, the claim that I was aiming at the pin, can be understood as the claim that there was a higher probability of my adjusting θ and v to within a relatively narrow range around the values that result in a hole in one, while H_2 is the claim that there was a reasonable probability of obtaining a broader range of initial conditions, centered on those which would land the ball on the green. Furthermore, all of the probabilities in play are objectively obtainable. We could, for instance, measure the frequency with which different values of θ and v obtain when I aim at the pin and swing and use these to estimate the probability distribution over possible initial conditions. We can thus determine in an objective fashion that $P(O \mid H_1) > P(O \mid H_2)$, and assert without prejudice that the observation O better supports the hypothesis that I was aiming at the hole.

What's important to note in this example is that, when we were worrying about the space of possibilities compatible with H_1 or H_2, we considered the range of initial conditions. The possible values of θ and v were not determined by bare logical consistency, but by the very practical constraints of what I actually do when I swing a golf club (that's why we could, for example, rule out head speeds greater than the speed of sound). What we did *not* worry about were the values of certain parameters in the laws we used to compute outcomes from initial conditions. That is, we did not ask about the probability of Newton's gravitational constant G having the value it does. To avoid giving the wrong impression, I do not mean to claim that it is necessarily a mistake to consider multiple possible values of G. I only mean to emphasize that parameters like G which occur in the laws linking initial conditions to outcomes are very different from the initial conditions of a process governed by those laws. It is usually reasonable to consider varying the initial conditions while keeping the laws fixed. Craig frequently blurs this distinction, speaking about the coupling constants that appear in the laws of particle physics as if they were the same sort of thing as "the initial conditions of the Big Bang."[20] The latter are presumably something like the distribution of matter and energy in the universe over some region of spacetime and are analogous to θ and v in my example. Since fine-tuning arguments will often ask us to consider variations in the parameters characterizing laws rather than just the initial conditions, it is important to keep the distinction in mind. One might accept the possibility of different initial conditions while rejecting the possibility of different parameter values. At the very least, one might think there are objective ways to attribute probabilities to initial conditions that do not work for the parameters of the laws themselves.

17.6 A likelihood argument from fine tuning

Robin Collins offers a design argument from fine tuning in the form of a likelihood argument.[21] It begins, as do all fine-tuning arguments, with the surprising observation that many of the parameters characterizing physical laws are fine-tuned for life. Collins then goes on to compute the likelihoods for various hypotheses in light of this observation, and ultimately favors design (or a multiverse, but we'll come to that in a moment). More so than either Craig or Davies, however, he tries to be explicit about what is meant by fine tuning and about how we should go about attributing probabilities to various outcomes under the hypothesis of chance. Let's take each of these issues in turn.

As Collins puts it:[22]

> As a first approximation, we can think of the claim that a parameter of physics is "fine-tuned" as the claim that the range of values, r, of the parameter that is life-permitting is very small compared with some non-arbitrarily chosen theoretically "possible" range of values R. The degree of fine tuning could then be defined as the ratio of the width of the life-permitting region to the comparison region.

This basic idea is depicted graphically in Figure 17.2. The horizontal axis represents all of the values that can, mathematically at least, be attributed to

---

[20] Craig 2003, 157.
[21] See, for instance, Collins 2003 and 2009.
[22] Collins 2003, 179.

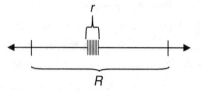

Figure 17.2   A depiction of the range of parameter values for which carbon-based life is possible compared to the theoretically possible range of parameter values (bracketed region of width R)

the parameter under consideration (e.g. the fine-structure constant α). The vertical tick marks indicate the range of "theoretically possible" values – the values which this parameter can have in any physically possible universe. Within that broad range is a very narrow range of values the parameter can take if carbon-based life is to be possible. This range is indicated by the striped region. The 'degree' of fine tuning, the extent to which we should be surprised that the parameter is such as to permit life is just the ratio $\frac{r}{R}$.

Collins goes on to refine this notion of fine tuning by pointing to other subtly different ways in which we might say that a parameter is fine-tuned. First, there is the notion of optimality. Collins thinks that, even if $\frac{r}{R}$ is not especially small, we might still want to say that the parameter in question is fine-tuned. We would do so if its value happens to lie within the small portion of r that is *optimal* for the production of carbon-based life. Even if the life-permitting region r is quite broad, there is likely only a small set of values within r that is optimal. Additionally, Collins thinks that we can speak of *one-sided tuning*. Suppose that both the "theoretically possible" values of the parameter and the life-permitting region are bounded on only one side. Then both R and r would be infinitely wide. For instance, we might know that decreasing the value of a parameter like the weak-force coupling constant would prohibit life, but we have no reason to believe that strengthening it to any degree would prohibit life. In that case, we can still look at where in the life-permitting range the actual value of our parameter falls. Collins thinks that it would be a remarkable coincidence if the actual value were very near the edge of r, in this case near its lower bound.

Both of these refinements introduce further controversial premises. The notion of optimality would seem to require that we conceive the setting of parameters as some sort of process with a natural notion of efficiency attached. If we attempt to understand 'optimal' as simply metaphorical for 'makes highly probable the emergence of life', then judgments of optimality require us to know a lot more about the emergence of life than is necessary to judge claims like "carbon-based life is possible." It is not clear that we possess anywhere near the requisite understanding, and so an argument based on optimality could only be viewed as hypothetical. On the other hand, the notion of a one-sided fine tuning requires us to set an arbitrary value for what counts as being 'near' the boundary of an infinitely extended region. Imagine that you are standing on a line that begins at some point but extends infinitely in the other direction (something like this: •→). How far from the initial point counts as close? Do you have to stand within an inch? A mile? A million miles? Whatever you pick, you are still infinitely far from the other end. Unlike the case of finite regions r and R, there is just no natural scale and no non-arbitrary way to set the boundary. Because of these obvious difficulties, I will stick with Collins' initial definition of fine tuning for the remainder of the discussion.

Before presenting his argument, there is one more important feature of Collins' proposal to emphasize. As he points out in the quotation given above, there must be some non-arbitrary way of specifying R. Collins' approach is to point to the range of values spanned by similar parameters. For example, suppose we are considering the possible values for a parameter corresponding to the strength of the strong force. Collins' approach is to note that, of the four known forces, the 'strengths' range from α_G (a dimensionless measure of the strength of gravity) to $10^{40}\,\alpha_G$ (the value of the strong-force coupling constant). Of course, it is a bit misleading to call α_G the 'strength' of gravity in any absolute sense. The parameter α_G is a coupling constant like the fine-structure constant, but with one important difference. There is no fundamental unit of mass. In this case, the parameter was computed using the mass of the neutron. The choice, however, is arbitrary. One could just as well choose to compute the parameter using my mass instead. In which case, its value would be different by many orders of magnitude. Furthermore, why should the values taken by different, independent parameters in the laws of physics put bounds on the possible values of the strength of the strong force? If in fact these parameters can be varied independently as Collins implicitly assumes, then it is difficult to see why the values taken by one parameter should tell us anything about the possible values of another. In the next chapter, we'll see what McGrew and colleagues make of this tactic. For now, let's grant that such an inference is justified and look at how it underwrites a likelihood argument.

The setup is by now quite familiar. Let O be the fact of fine tuning in Collins' sense. That is,

> O: One or more parameters in one or more physical laws are fine-tuned – they lie in a relatively narrow range of life-permitting values in a broad range of theoretically possible values.

We can consider the following hypotheses to explain this fact:

> H_1: The relevant parameters acquired their values by chance and there is only one universe.
>
> H_2: The relevant parameters acquired their values by design and there is only one universe.
>
> H_3: The relevant parameters acquired their values by chance and there are a great many (perhaps infinitely many) universes.

To conclude the likelihood argument we compute the relevant likelihoods. In particular, in order to compute $P(O \mid H_1)$, we assume a uniform distribution over the values in R, so that $P(O \mid H_1) = \frac{r}{R}$. Then, for any particular instance of fine tuning singled out by Collins, we can say that $P(O \mid H_1) \ll P(O \mid H_2)$ and $P(O \mid H_1) \ll P(O \mid H_3)$. The latter inequality follows because after enough random 'draws' even improbable events become plausibly probable. Thus we should favor both the design hypothesis and the multiverse hypothesis over chance, though it is difficult to say which of these two to prefer on likelihood grounds alone. In later work, Collins argues against the sorts of multiverse hypotheses I discussed above by pointing out that they all presume some sort of physical mechanism by which universes are produced with different values for their parameters. Whatever that physical mechanism might be, it simply invites another fine-tuning problem: why is the mechanism just right for making life-sustaining universes probable? Because it ultimately fails to explain fine tuning, Collins urges us to reject the multiverse hypothesis.

18 Fine tuning II: objections

18.1 Introduction

Design arguments from fine tuning come in a wide variety of flavors, but they all share some common ingredients. For starters, they all rest on the observation that the dimensionless parameters characterizing the laws of physics are 'fine-tuned' for life. Were we to dial in ever-so-slightly different values, so the story goes, we would find ourselves with a vastly different universe inhospitable to the likes of us. This fact of fine tuning cries out "conspiracy!" That's a fine intuition, but an argument needs to make a rational case for such a claim. The typical fine-tuning argument makes this case in terms of probability or a related measure of how unlikely such a configuration would be given that the parameter values are the product of unintelligent chance. The result is overwhelming confidence in the design hypothesis.

Every element of the typical fine-tuning argument has been called into question. Some take issue with the claims of fine tuning themselves, arguing that the laws are not so finely adjusted and thus there is nothing to explain. More mathematically minded objectors have challenged the use of probabilities in those forms of fine-tuning argument that appeal to likelihoods or specified complexity. Finally, there are philosophical objections that take issue with the metaphysics implicit in describing a problem of fine tuning in the first place. In order to see a puzzle, in order to recognize a potential conspiracy, the advocate of fine tuning requires us to adopt a package of metaphysical positions on the nature of physical laws and our knowledge of them. Under a family of Humean views of science, there is no fine-tuning problem. We'll take up each of these sorts of objections below. In the next chapter, we'll take a step back and see where our ride through two millennia of design arguments has landed us. The answer suggests that attempting to untangle our intuitions of design may yet have something vital to teach us about the universe, irrespective of the ultimate verdict on design arguments.

18.2 Not so finely tuned

As we saw in the last chapter, Carr and Rees argue that much of the structure of the universe is determined by the values of a handful of "fundamental constants."[1] In particular, features essential to the existence of carbon-based life like us seem to depend rather delicately on combinations of the fundamental parameters. Consequently, if any one of those dimensionless parameters was to have a slightly different value from that which it is actually found to possess, living things could not exist. This theme has been greatly elaborated since that seminal paper, but the basic claim is at the root of fine-tuning arguments. These are the facts which suggest to some that the laws of physics have been intentionally adjusted to allow for life like us to flourish.

But are these facts? Robert Klee argues that Carr and Rees and many other fine-tuners have overstated the coincidences that drive our intuitions of fine tuning.[2] For instance, Carr and Rees suggest that unless the gravitational fine-structure constant, α_G, is approximately equal to the twentieth power of the fine-structure constant, α (a fact we denote by $\alpha_G \sim \alpha^{20}$), then it would not be possible to form the kind of stars that can have planetary systems. As they put it, if α_G were "slightly larger, all stars would be blue giants; if it were slightly smaller, all stars would be red dwarfs."[3] The significance of this fact is that blue giants are supposedly inhospitable to planetary systems while red dwarfs do not explode to spread elements like carbon about the universe. Whether or not this is so, let's focus on the factual claim. Is it true that $\alpha_G \sim \alpha^{20}$, and if so, how much is "slightly" more or less? As Klee points out, α_G has a value of 5.90499×10^{-39}, whereas α^{20} $= 1.8336 \times 10^{-43}$. In other words, α_G is 30,000 times bigger than α^{20}! It is thus a little hard to accept that they are approximately equivalent. This is not an isolated example, either. Klee provides a table showing computed values for all of the coincidences cited by Carr and Rees. The accuracy of

---

[1] Carr and Rees 1979.   [2] Klee 2002.
[3] Carr and Rees 1979, 611.

approximate equality in these relations can be assessed by taking the ratio of one quantity to the other (i.e., x/y). Of the thirteen remaining relations presented by Carr and Rees, for only two is it the case that $1 > x/y > 0.98$. For the remainder, one quantity is at least 1.85 times bigger than the other, and in many cases more than ten times bigger. According to Klee, similar problems arise throughout the fine-tuning literature, and egregious examples are not hard to find.

What does this mean for fine-tuning arguments? What Klee's results indicate is that Carr and Rees oversold the claim that a handful of dimensionless constants precisely determine all of the "natural scales." It means that other facts, such as the specific, contingent arrangement of stuff in the early universe, must have a lot to do with the structure we see – it isn't all determined by the parameters in the laws of physics, but rather in part by the initial conditions. Yet one might still have the makings of a fine-tuning argument, even if the coincidences cited between dimensionless parameters aren't precise. One could still argue that had the combination of parameters and initial conditions been slightly different, then the universe would be bereft of life. That means that what we really need to know is how sensitive this combination is with respect to the possibility of life. Just how much can we vary things and still get the features that fine-tuning arguments rest upon, such as life and complexity? If we are to construct a fine-tuning argument the answer must be that we could only change things ever so slightly without ruining the life-supporting capacity of the universe. So how slight is "slightly"?

For at least one famous example, suggests Victor Stenger, the answer is a lot. That is, there is quite a significant range over which we can vary the relevant parameter and still get a life-friendly cosmos. The example in question is the 'Hoyle resonance'. All of the elements in the universe heavier than helium – including carbon, the basis of life as we know it – are formed in the nuclear furnaces we call stars. In the case of carbon, the synthetic process involves a two-step dance. First, two helium nuclei fuse to form beryllium. Second, a beryllium nucleus fuses with an additional helium nucleus to form carbon. But the latter can only happen fast enough to produce the amount of carbon we find in the universe if the carbon nucleus has an "excited state" with an energy close to that which the heat of a star provides to colliding helium and beryllium nuclei. For our purposes, it's enough to think of the excited state of carbon as a configuration of the carbon nucleus that makes it look more like the state of a beryllium and an energetic helium nucleus together. This makes the probability of the reaction much higher than if there was no excited state. So Hoyle made a prediction based on 'anthropic' reasoning: if we see this much carbon in the universe, all of that carbon is made in stars, and the only way that could be so is if carbon nuclei have an excited state of the right energy (7.7 MeV), then there must be such an excited state. This prediction was quickly confirmed experimentally, and has since stood as the principal (if not the only) example of successful anthropic reasoning.

It is also the sort of coincidence that fine-tuning arguments rest upon. The idea is that even a slightly different energy for the excited state of carbon would mean that the universe would fail to produce enough and so life would be impossible. However, Stenger points to the work of physicist Mario Livio and his collaborators to argue that there is really quite a wide range of energies at which the carbon excited state could occur and still allow for enough carbon to be produced.[4] Specifically, Livio *et al.* found that the excited state could occur at an energy level higher or lower than that observed by 0.06 MeV with no appreciable difference in the amount of carbon produced. Of course, we may not need the exact amount of carbon we have for life to be possible. We could plausibly make do with significantly less or significantly more carbon. In that case, we could alter the location of the excited state energy by 0.277 MeV before life becomes untenable.

But is this really a problem for those who claim that the resonance is fine-tuned? One might reasonably respond by saying that a range of 0.12 MeV doesn't seem so big. Livio *et al.* put their finger on precisely this issue: "The implications for these results for evaluating the anthropic 'coincidence' on which our existence seems to rely are not entirely free from subjective feelings. Whether or not one must conclude that the [excited state] has to be exactly where it is depends on whether we should regard a change of 60 keV as small or large."[5] As we'll see in the next section, this ambiguity over what counts as finely tuned – the problem of how small is small – is deeply problematic for fine-tuning arguments that invoke probabilities.

[4] Livio *et al.* 1989.   [5] Livio *et al.* 1989, 284.

18.3 Problems with probability: ill-defined probabilities

In all of the arguments from fine tuning we've looked at, a crucial premise is that the probability of attaining the actual parameter values by chance is vanishingly small. In other words, it is assumed that there is a rigorous way to apply a probability measure to a range of parameter values and so determine that the ones we have are improbable. A number of authors have challenged this assumption. I'll focus on one particular skeptical analysis advanced by Timothy McGrew, Lydia McGrew, and Eric Vestrup.[6] They argue that there is no non-arbitrary way to define a well-behaved probability distribution over the range of possible values for any given parameter. To see the force of their argument, we'll proceed in stages. In the first stage, we'll consider an example from the philosopher John Leslie for which there *is* a well-defined probability measure over a continuous space of possibilities (by a continuous space I just mean something with an infinite number of points infinitely close together like a Euclidean plane or the real-number line). Then in the next section we'll consider the problem of specifying the space of possible values for a parameter, and see how this affects our ability to use the probility calculus. Finally, we'll consider a possible rescue of the fine-tuning argument and the rebuttal provided by McGrew *et al.*

As you may recall from Chapter 11, probability theory begins with the specification of the collection of possible outcomes for a given random process. This collection is called an 'event space'. Probability is like paint. You have a unit amount of it with which you can paint each of the events in the event space. Put another way, the probability of an event is a number between 0 and 1 and such that the sum of the probabilities of all possible events is 1. Now, aside from the probability of events in the event space, you can also consider the probability of sets of events as well as the unions and intersections of these sets. The union of any number of sets must, according to the rules of probability, amount to no more than 1.

The fact that we can consider sets of events rather than individual events is helpful, particularly in cases where there are infinitely many events in the event space. Consider, for instance, a dartboard.[7] The face of the dartboard is a disc of finite (and pretty modest) width. In its mathematical abstraction, the disc of the dartboard is made up of infinitely many

---

[6] McGrew *et al.* 2003.   [7] See Leslie 1989, 199–200.

points. Nonetheless, we can assign probabilities to sets of these points – regions of the dartboard – such that the union of all sets is the whole dartboard and the probability assigned to the whole dartboard is one. With such a probability measure, we can sensibly ask for the probability that a dart thrown at random strikes any given region of the board such as the bull's-eye. It doesn't matter that there are infinitely many points the dart could strike.

Now, if we are modeling the actions of a very poor dart player, we might want to claim that the probability of striking any region of the board with a given area is the same as striking any other region of the same area. We would call this a 'uniform' distribution. With a uniform distribution, every place on the dartboard is as likely to be hit with a dart as any other. This sort of distribution is possible for the dartboard because no matter what area we pick, we can divide the board up into a finite number of regions of that area. So for instance, we could cut the dartboard into pieces each of which is 1 cm^2 and we will end up with a finite number of pieces. This is important, because probability theory demands that any well-behaved probability measure have the following property: the sum of the probabilities of all the disjoint pieces must add up to 1. This is isn't surprising because we said the probability of hitting the whole dartboard (hitting the dartboard anywhere) is unity, and the pieces combine together to make up the whole dartboard.

Where we run into trouble is when our dartboard is infinitely large. In that case, there is no way to assign the same probability (which is a positive number) to every piece of equal size and still have the probabilities of the whole board add up to 1. Infinity times any positive number is infinity! According to McGrew et al., this is the problem faced by the proponent of the fine-tuning argument. Suppose we are worried about the fine-structure constant, α. The reason this parameter is of interest is precisely that its value is not a consequence of any known physical theory. It is a number we have to put in and thus the sort of thing you could 'tune' in order to adjust the laws of physics and the contents of the universe they produce. From this perspective, the only constraint on values that α can take is that it is a real-valued parameter – any real number from negative to positive infinity is a permissible value. But there cannot be a uniform probability distribution over this space of possibilities. We cannot claim that every segment of the real line that is the same size should receive the same non-zero probability, since any such probability we try to assign will not sum to unity for the entire space of possibilities. Thus, we cannot speak of the 'probability' that a randomly selected value for α is within the range that is life-permitting. To speak this way is gibberish.

You might think all this worrying is for nothing. After all, we might adopt Collins' strategy and just assume that there is only a finite range of possible values for any given parameter. Then the situation is the same as for a dart-board, and we can give well-defined probabilities. If we could specify a min-imum and maximum possible value for α, then we would be able to assign a uniform probability distribution over the space of possible values.

But as McGrew et al. point out, there doesn't appear to be any non-arbitrary way of deciding on what those upper and lower bounds are with-out also begging the question. As I mentioned in the last chapter, Collins uses the values of other parameters to set bounds on any given parameter like α. But why should the values actually taken by some other parameter tell us anything at all about the possible values of the one we're interested in? As McGrew et al. put it, "There is no *logical* restriction on the strength of the strong nuclear force, the speed of light, or the other parameters in the upward direction."[8] So it seems we have to consider an infinite range of values if logical consistency is our only guide. Any choice of a finite range would seem to beg the question because the specific choice determines whether life-permitting values are probable or not. Since there are no phys-ical constraints on the possible values, logical consistency is all we have to go on.

There is one more way out that we have not yet considered. We might embrace the claim that the range of possible values for a parameter is infin-ite, but reject the claim that our ignorance must correspond to a uniform probability distribution. How could this help? Think of it this way. If we try to give the same probability p to each segment of length 1 in the line extending from 0 to positive infinity, then the probability associated with the whole line diverges to infinity. This is because the sum $p + p + p + \ldots$ does not converge on a finite number as long as $p > 0$. But suppose instead that we attributed smaller and smaller probabilities to each of the pieces. The sum

$$(1{-}p) + (1{-}p)p + (1{-}p)p^2 + (1{-}p)p^3 + \ldots$$

8 McGrew *et al.* 2003, 201; emphasis in original.

does in fact converge (or sum up in the limit) to 1. So if we assigned the probabilities in this series to each successive segment, we would have a proper probability measure!

The sort of probability measure we're talking about is called a normalizable continuous distribution. An example is the bell curve you find everywhere in statistics. The problem is that there are infinitely many such normalizable distributions and no grounds for choosing one over the other. Any one distribution makes certain values of the parameter more likely than others. But "[w]orking from bare logical possibilities, it seems unreasonable to suggest that any one range of values for the constants is more probable *a priori* than any other similar range – we have no right to assume that one sort of universe is more probable *a priori* than any other sort."[9] The only justifiable distribution seems to be the uniform distribution. But the uniform distribution cannot give a viable probability measure! Thus we are stuck with an impossible set of demands to satisfy. It seems that the proponent of fine tuning cannot appeal to the probability calculus to make his intuition precise. At least, he cannot do so without justifying on purely logical grounds either a finite range of parameter values or a non-uniform distribution.

18.4 Problems with probability: observational selection effects

I want to turn now to a consideration of what Elliott Sober calls an 'Observational Selection Effect' (OSE).[10] Imagine that some fishermen return from fishing at a lake and report a surprising fact:

O₁: All fifty of the fish caught in the lake were greater than 10 inches in length.

There are obviously a great many hypotheses we might consider that could account for this fact, but Sober asks us to consider two:

H₁: All of the fish in the lake are longer than 10 inches.
H₂: Half of the fish in the lake are longer than 10 inches.

⁹ McGrew *et al.* 2003, 204.   ¹⁰ Sober 2003.

Appealing to the Likelihood Principle, we note that $P(O_1 \mid H_1) > P(O_1 \mid H_2)$ and so we should favor the hypothesis that all the fish in the lake are longer than 10 inches. Now suppose, however, that the fishermen go on to explain *how* they came by their catch: they have been trawling with a net with 10-inch holes. Let's call this fact A_1. Now, intuitively this extra information ought to change our assessment of the likelihoods. But how? One response is to insist that one append to H_1 and H_2 all relevant background information. This is something like a 'Principle of Total Evidence'. When we do this and recompute the likelihoods, we find that

$$(O_1 \mid H_1 \,\&\, A_1) = P(O_1 \mid H_2 \,\&\, A_1) = 1.$$

Because the fact of A_1 makes the observation O_1 necessary – since fishing with such a net means necessarily catching only fish bigger than 10 inches long – the details of our hypotheses are irrelevant. The probability of catching only fish greater than 10 inches long is unity.

The impact of the additional fact A_1 on the computed likelihoods is an example of an OSE. This notion is captured in the following general definition:

OSE: The alteration of likelihoods involving an observation fact due to the inclusion of information about how that observation fact was obtained constitutes an *observational selection effect*.

Note that OSEs needn't be as dramatic as in the fishing example. Information concerning how observations were obtained may alter the likelihoods but not reduce them all to unity. For example, suppose we were to discover that an overwhelming fraction of people living in a rural region of China support the damming of a river that would submerge much of the arable land in the area. Let's call this O_2. We might hypothesize that this support might be explained by any number of hypotheses H_i, with corresponding likelihoods $P(O_2 \mid H_i)$. But suppose that we come to learn that the observation was acquired by taking an online poll which required Internet access, and the only people in the region with Internet access are developers or local industries who will profit from the dam. This additional fact A_2 will raise all of the likelihoods. For any given hypothesis, $P(O_2 \mid H_i \,\&\, A_2) > P(O_2 \mid H_i)$ simply because A_2 implies that the only people surveyed tend to be in favor of the project – we did not take a random sample of the population.

To see how OSEs might be relevant to fine-tuning arguments that invoke likelihoods (as Craig's argument does), it will help to consider a second example taken up by Sober. Originally due to John Leslie,[11] this is the infamous firing-squad example. To personalize it a bit, suppose that I am to be executed by firing squad. I am blindfolded and stood before a wall while twelve sharpshooters take aim and fire twelve rounds each. Remarkably, when the din ceases and the smoke settles I make the following remarkable observation:

O_2: Ben is still alive.

Now, there are two obvious hypotheses we might advance to account for this fact:

H_3: The sharpshooters all happened to miss by chance.
H_4: The sharpshooters intended to miss.

We can understand these hypotheses by supposing that, when a sharpshooter intends to shoot someone, there is actually a distribution of possible initial conditions for the bullet, most of which result in the bullet striking the prisoner. H_3 is just the claim that all twelve sharpshooters for all twelve rounds happened to randomly instantiate the tiny fraction of initial conditions open to them that lead to a miss. Obviously, this is rather improbable. On the other hand, when he intends to miss, presumably there is only a tiny probability that the initial conditions instantiated by a sharpshooter pulling the trigger will result in the death of the prisoner. With these facts in mind, facts that should strike you as being very similar in kind to those brought to bear in my golf example of the last chapter, we can determine that

$$P(O_2 \mid H_4) \gg P(O_2 \mid H_3).$$

Thus at first glance we should favor the hypothesis H_4. However, this is to ignore some relevant information about how we came by the observation. If you happened to be in the crowd and observed that I was still alive after the squad stopped shooting, then you would be justified in conducting the analysis as I've presented it so far. You could be pretty sure that the

[11] Leslie 1989.

sharpshooters meant to miss. However, if I made the observation that I am still alive (as was the case given the way I framed the story) then I have to take this fact into account:

A$_2$: The fact that Ben is still alive was ascertained by Ben.

But in this case, $P(O_2 \mid H_4 \& A_2) = P(O_2 \mid H_3 \& A_2) = 1$. I, as the observer, have no grounds for favoring one hypothesis over the other! The very fact that I made the observation entails the truth of the observation. Irrespective of which hypothesis I consider, the probability of O_2 is always unity.

What do OSEs have to do with fine tuning? When speaking about the universe in which we live being fine-tuned for life, we are very much in the position of the prisoner observing that he is still alive. To make a fine-tuning argument by appealing to the Likelihood Principle, one compares a couple of probabilities. Let E stand for the observed fact that the parameters of physics are fine-tuned for life, let H stand for the chance hypothesis, and let D represent the hypothesis of design. It seems that $P(E \mid D) \gg P(E \mid H)$. The problem is that like the survivor, we collectively suffer from a strong OSE. The observation that E is the case was obtained by carbon-based life forms. If we take this fact, call it A$_3$, into account then all likelihoods reduce to unity: $P(E \mid H \& A_3) = P(E \mid D \& A_3) = 1$. The mere fact that we observe E to be the case entails that E must be the case. Life forms like us wouldn't be here to observe E unless E were true. Thus, when the method by which our observations are acquired is taken into account, the justification for preferring design to chance (or any other hypothesis, for that matter) vanishes.

Sober notes that the proponent of fine tuning might attempt to escape this objection by speaking not of likelihoods but of probabilities. That is, one might compare the *probability* of the design hypothesis given the evidence: $P(H \mid E)$. As you may recall from our discussion of Bayes' Theorem, this probability is given by:

$$P(H|E) = \frac{P(E \mid H)P(H)}{P(E)}$$

But to attempt this maneuver, says Sober, is to invite a fresh critique. What could it mean to attach a firm probability to the evidence at hand? How could one objectively evaluate $P(E)$? This is not the same sort of well-posed problem that we faced when putting probabilities over initial conditions for my golf swing. It is not clear how we can make sense of what this sort of probability represents. Furthermore, even if there is a way to make sense of $P(E)$, how do we go about determining what the distribution is over possible ways the universe might have been, particularly if we have only one universe?

The defender of fine tuning needn't go down this difficult road, however, since there is something very wrong with the OSE objection. The first clue that something is amiss was given to us in the firing-squad example. There, the very same evidence gave very different assessments of the given hypotheses depending upon who was doing the analysis, someone in the crowd or the person being shot at. If the likelihood approach to hypothesis assessment is objective, our judgments about hypotheses ought not to depend on who's asking the question! Fortunately, there is a straightforward way to untangle this knot without abandoning the Likelihood Principle which Sober favors. The trick is to distinguish between two distinct questions. On the one hand, we might want to know which hypothesis is favored by a given piece of evidence in the context of all the facts we already have in hand. Let's suppose we represent all of the things we already know by B. I mentioned in Chapter 11 one of the main considerations that make the Likelihood Principle plausible in the first place, namely that a particular ratio of likelihoods does all of the work in shifting our posterior odds in favor of a hypothesis. Now, the prior probability of hypothesis H – whether we know it or not – is given by $P(H \mid B)$. So the same considerations that lead us to adopt LP suggest that this first sort of question is answered by computing this likelihood: $P(E \mid H \& B)$. If that's all there was to it, then Sober would be right and fine-tuning arguments would face serious problems with OSEs. However, there is a second sort of question we might be asking: which hypothesis does the *totality* of evidence in hand favor? In other words, relative to a state of total ignorance, does all the evidence we have favor H_1 over H_2? The same sort of argument used to justify LP suggests that this question is answered by computing $P(E \& B \mid H_1)$ and $P(E \& B \mid H_2)$. These likelihoods might still be different even if B implies E.

Let me illustrate the distinction by returning to the firing-squad example. Given that I already know A_2, namely that the observation of my survival was made by me, the additional piece of evidence, O_2, that Ben is still alive does not weigh in favor of either hypothesis, since $P(O_2 \mid H_4 \& A_2)$

$= P(O_2 \mid H_3 \& A_2) = 1$. However, we might reasonably ask whether *all* of the evidence together favors one hypothesis over another. That is, do the facts that I observed my survival and the fact that Ben is still alive together favor one hypothesis over another? To answer this we have to compute $P(O_2 \& A_2 \mid H_3)$ and $P(O_2 \& A_2 \mid H_4)$. Since it is overwhelmingly improbable that the sharpshooters all missed by accident, we can safely say that $P(O_2 \& A_2 \mid H_4)$ $\gg P(O_2 \& A_2 \mid H_3)$. So the OSE is irrelevant when determining the impact of the overall evidence. This accords well with our intuitions – of course my survival favors the conspiracy hypothesis. What does this mean for fine-tuning arguments? While it may be completely uninformative to discover that the laws of physics are just such as to allow for carbon-based life like us given that we already know we're here to make the observations, the fact that we are here at all may weigh in favor of the design hypothesis over that of chance. In other words, it may very well be that $P(E \& A_3 \mid D)$ $\gg P(E \& A_3 \mid H)$. OSEs are not a problem for fine-tuning arguments.

18.5 What are the possibilities?

Whether or not we try to use the probability calculus, all fine-tuning arguments depend upon claims of surprisingness – how astonishing it is that
the universe is just right for life. This sense of surprise in turn depends on
a contrasting set of possibilities. If, in some sense, the universe could only
have been the one way in which we find it, then it isn't surprising that's
the way we find it. Rather, those who see fine tuning in the laws of physics
suppose that there is an enormous variety of ways in which the universe
might have been. It is only because a tiny fraction of those ways would
have supported life that we find the life-supporting values of physical
parameters surprising. But how can we justify the claim that many different universes are possible? Precisely which ones are possible, and how rare
are those that contain something we might call life? In the last section, we
examined problems that follow from assuming that any given parameter
could take on any real number value. In particular, we found it extremely
difficult to choose a range

But there is another point to be made here. What matters for the possibility of life is not any one parameter. What matters is the full set of
parameters and how they coordinate to determine the physical contents of
the universe. And we expect there to be significant coordination, whether

or not the laws are products of design. None of the relations cited by Carr and Rees showing how a given life-supporting phenomenon is determined by the values of a few fundamental parameters is all that surprising when taken in isolation. At least, we ought not to be surprised that complicated physical phenomena like stars are addressed by multiple branches of physical theory like gravity and electromagnetism. If there was no such overlap amongst physical theories, all phenomena would be divided into isolated sets by the physical laws – nothing involving electricity could ever have anything to do with a gravitational phenomenon. Such a world would probably be devoid of life if such a world is even conceivable, since living things involve interaction between mechanical and electric forces, thermal and chemical forms of energy, etc. So we should expect the laws of physics to overlap, and thus we should expect relations of the sort Carr and Rees emphasized.[12] But these relations tell us that changes in one parameter can be compensated for by changes in other parameters. Suppose we consider star formation, one of the examples I described in the last chapter. Increasing the parameter α_G while keeping the other parameters fixed would give a universe with only blue giant stars (and, presumably, no life-sustaining planets). However, if we also adjust α such that the ratio of the two parameters stays roughly the same, then we would still find stars like ours. The point is that a procedure in which we hold the values of all other parameters fixed while varying the one of interest is one that imposes an artificial restriction, one that overestimates the fragility of physical phenomena. What we need to know for a successful fine-tuning argument is just how fragile the whole network of relations amongst the physical parameters is. How much (and in how many ways) can all the parameters be varied together and still yield a universe like ours?

At least some authors have attempted to address this question. Victor Stenger has built a simplified model involving four constants: the fine-structure constant, α, the strong nuclear force strength, α_S, the mass of the electron, m_e, and the mass of the proton, m_p.[13] These constants are used to compute at least rough estimates for quantities like the radius of the hydrogen atom, the lifetime of stars like the Sun, and the minimum and maximum masses of planets. He then randomly varies the four parameters

---

[12] Cory Juhl (2006) offers a critique of design arguments based on similar considerations.
[13] See Stenger 2011, ch. 13.

over ten orders of magnitude, assuming that each order of magnitude is equally likely (in other words, he assumes a uniform distribution over the logarithm of the values of the parameters). He finds that in a surprisingly large fraction of cases, each of the quantities necessary for life is in fact realized. In other words, the physics doesn't look very fine-tuned at all. I won't go into the details of Stenger's simulation. As interesting as the results are, Stenger's model raises as many questions as it answers. For instance, why restrict the values of the parameters to that particular range? Why assume an even distribution over a logarithmic scale (i.e. why this particular choice of distributions)? Why not take an even distribution on an ordinary linear scale? These are all questions we posed for the defender of a fine-tuning argument, and they are questions that must be addressed if one wishes to deny claims of fine tuning in this way.

The ultimate worry for fine-tuning arguments is that there is no good way to pick out a space of possibilities. If there is a physical theory that determines the possibilities, then this suggests that there may be a physical reason for the universe to have its life-sustaining properties, and thus no room for fine tuning. If there is no such physical theory, then we are left seeking a metaphysical one that is independently justifiable and neutral with respect to the question of design. No such theory has been proposed other than the very weak restriction of conceivability. That is, we are left assuming that the possible universes are whatever ones we (or some idealized thinker) can coherently imagine. But this just opens up other problems. One might argue, for instance, that even those who consider varying all of the parameters simultaneously have not gone far enough. It is conceivable that the laws of physics could have been different in other ways rather than the values of their parameters. One could even make the case that there is no reason to distinguish between parameters and the 'form' of physical laws. If you change the value of a parameter, then you select a distinct function. So why not choose from a broader class of functions? Why not, for instance, vary the exponent in Newton's law of universal gravitation so that the strength of gravity varies inversely as the distance cubed? There seems to be no argument – at least an explicit argument is hard to find – as to why we should not consider alternate forms of the laws in addition to different parameter values. One reason participants in this debate likely shy away from this question is that we don't know how to entertain this possibility in any concrete way. It's challenging enough to specify a plausible space of parameter values. The intuition seems to be that it would be hopeless to attempt to characterize all possible forms of the laws.

The inability to justify a choice of possible universes remains deeply problematic. It is hard to make the case that we should be surprised about the universe we have when there are no good reasons to believe in one or another set of alternatives. I should note that this problem is not shared by those older cosmic design arguments that appeal to configurations of matter that are unusual or inexplicable within this physical universe and with respect to the physical laws we have. The argument from Newton and Bentley about planetary orbits didn't have this problem because eventually nature or physics will tell us which initial conditions are possible in the formation of a solar system and even how frequent each such condition is. That's exactly what we're learning as we discover increasingly vast numbers of extrasolar planetary systems. There is no analogous source of information in the fine-tuning case. As Hume said, we don't get to observe the creation of universes. Instead we are left with a dilemma. Without a compelling theory of what laws are possible, fine-tuning arguments cannot get off the ground. At the same time, it seems we need to deny the possibility of such a theory in order to implicate design.

18.6 The big picture: arguing for the need to explain

Whether cast as an argument from order or a likelihood argument, arguments from fine tuning all rest on a supposition that there is something about the laws of physics that requires an explanation. You might wonder why anyone would worry about such an issue. Surely almost any fact is something for which we can reasonably demand an explanation. But this case is special. When we ask about the form of the laws of physics, we are not asking about a regular sort of occurrence within the universe. Asking why Newton's G has the value it does is not the same as asking why the engine in my truck is idling rough. The latter sort of question can be answered by appealing to other facts about the world *and to the laws that govern such facts*. But this sort of appeal makes no sense when we ask about the laws themselves. Think of it this way. When learning how to play chess, it is perfectly reasonable to ask why a certain configuration of pieces results in checkmate. But your instructor is likely to give you a funny look if you ask *why* a pawn can only move one space at a time. The first question is answered by appeal to the rules governing play. The second question asks about those rules, and so no appeal can be made to them. Naturally, when speaking of human games like chess, there are other levels of explanation to which we can appeal. For instance, we could talk about human desires to produce fair games and what sorts of rules constitute a fair game, etc. But when asking about the rules which govern the entire universe, humans and all, it is not clear that there is another level of analysis to appeal to. In this sense, asserting that the form of the laws demands explanation is a claim that requires clarification and justification. We need to know why there is anything to explain, and, assuming there is, what sorts of explanations would be satisfying.

Of the authors considered above, only Paul Davies takes up the first concern explicitly.[14] Recall that the first part of his argument (see Chapter 17) constitutes an argument in favor of the claim that the laws demand explanation.[15] The argument went something like this:

(1) The physical laws of the universe are such as permit life.
(2) Other physical laws are possible.
(3) That physical laws are life-permitting is a contingent fact (from (1) and (2)).
(4) Contingent facts must have an explanation (contingent states of affairs have a cause).
(5) Therefore, the fact that the physical laws are life-permitting must have an explanation.

Let's grant that this argument is sound. The question then remains, what sort of explanation does the contingent fact of the laws having the form they do require?

18.7 The big picture: what sort of explanation?

There are a number of possible sorts of explanation. First, we might consider explanations in terms of other, contingent physical facts. For instance,

[14] Davies 2003.
[15] I am not being very careful to distinguish explanations from causes. This is sloppiness on my part. Explanation refers to a relation between propositions. Causation refers to a relation between events or states of affairs. I am willing to blur the distinction a bit because I assume that any suitable explanation of a proposition will refer to the state of affairs which is the cause of the state of affairs referred to in the proposition.

the parameters might have the values they do because some set of meta-laws or a grand unified theory of physics (the one law of unlimited domain that determines all the other laws of limited domain) requires this to be the case. In other words, it may be physically necessary for the laws to be as they are. But Collins, for one, explicitly denies that any such explanation would be satisfactory: "Even if such a theory were developed, it would still be a huge coincidence that the Grand Unified Theory implied just those values of these parameters of physics that are life-permitting, instead of some other value."[16] Collins seems to reject any explanation in terms of facts that are themselves contingent, that might themselves have been some other way. Since there are presumably many logically consistent unified theories of physics possible, any such unified theory would be contingent, and thus not satisfying as an explanation.

This brings us, then, to 'chance'. Many of the authors we have considered have at least entertained chance as a viable explanation for *some* contingent facts, if not the fact of fine tuning. Of course, for any particular contingent fact, the implicit claim is that chance is a satisfying explanation just if the chance process under consideration would make the fact highly probable. But whether it makes an event 'probable' or not, what does chance amount to as an explanation? All three authors speak of a chance explanation as if the fact in question – the fact of fine tuning – might have been randomly drawn from a pool of possible facts. What could this mean exactly? In what sense is a draw random when we are talking about a process that cannot be repeated even in principle? Perhaps more troubling is that referring to a fact as an outcome of chance seems to deny a cause or explanation altogether. Chance is the absence of a specific cause. When asked why x was chosen in a random draw, one could only reply "no reason – it might just have well have been y." Thus, accepting chance as a plausible candidate for explaining contingent facts would contradict the premise that every contingent fact has a cause. A chance event is an uncaused event.

This does not mean that design is the only possibility remaining, because it is not clear that design is a possibility either. That is, we might reject design as a plausible explanation on the very same grounds that we ruled out unified theories of physics. To attribute the form of the laws or values of the parameters to design is to posit an intelligence that exists outside of

---

[16] Collins 2003, 191.

the world governed by those laws, an intelligence that would be inclined to produce a universe with life in it. But many intelligences could be conceived which had the power but no such goals. That an intelligence willed the constants to have the values they do would itself be a contingent fact. And we already ruled out other contingent facts as plausible explanations.

Upon closer examination, then, it is simply not clear what sort of explanation or cause would constitute a valid answer to the question of fine tuning. It is difficult to tell whether an argument that favors design as an explanation is a strong one if we don't know whether design is even an explanation. While a lack of clarity of this sort does not mean a sound fine-tuning argument is impossible, it does make the soundness of those on hand doubtful. It is not that we have yet to determine whether the premises concerning the causes of fine tuning are true. Rather, we aren't sure how to decide whether the purported conclusion is coherent.

18.8 The big picture: metaphysical commitments

Rather than ask whether an explanation is required for the laws of physics, we can step back and ask what must be the case if the following is to be a coherent question at all: "Why are the laws of physics the way they are?" In the first place, we must accept that the laws of physics are as described by the Necessitarian, not the Regularity view. Were we to take a Regularity view of laws, the only explanations available to us for any given values of the parameters would refer to the facts about events in the universe. In the Regularity view, the laws are true generalizations about the physical events that make up existence. Thus, the parameters have the values they do because the facts are the way they are and because we choose to pay attention to some particular regularities amongst those facts. It is only under the Necessitarian view that we can sensibly ask which values of the parameters would cause the world to produce intelligent life. Only in the Necessitarian view are laws things that exist in their own right and thus have features that require their own explanations or causes.

Additionally, asking this question forces us to take a particular view of what we mean by 'universe'. One might think that the universe is by definition all that there is. Clearly, we cannot embrace this understanding while simultaneously talking about the causes of the universe, or the laws that govern the form of the universe. Rather, we have to have in mind by 'universe' something like the totality of spacetime events. Of course, this isn't really a metaphysical principle we're embracing, but rather a semantic decision – the stipulation of a definition. Nonetheless, it is an important point to keep in mind, since it is easy to conflate the two meanings.

More substantively, we must accept the existence of a mysterious relation of dependence. Whatever the relation is between contingent facts about things *outside* the universe (e.g. the will of a designer) and facts about things in the universe (e.g. the values of the constants), it cannot be the ordinary sorts of causation we are used to invoking. The latter only makes sense for things within the universe. That isn't to say that an account cannot be given of what such dependency would involve, only that it would have to rely on a priori metaphysics.

The purpose of reviewing these various commitments is not to motivate dismissing the question of fine tuning, but rather to clarify the sorts of commitment that come along with answering it. Many who dismiss the conclusion of design nonetheless embrace the question of fine tuning (e.g. the theoretical physicist Lee Smolin).[17] In doing so, they implicitly embrace some very strong views about the nature of the world, of natural laws, and of metaphysical possibility. Given some of the arguments such critics level against the design hypothesis, it is not always clear that they are aware of these commitments. If philosophy is about anything, it is about getting clear on one's assumptions.

[17] Smolin 1997.

19 Conclusion

19.1 Recapitulation

Most of the design arguments appearing in this book can be viewed as variations on a handful of themes already present in Stoic thought by the first century BCE. I've been calling these the argument from order, the argument from purpose, the argument from providence, and the argument by analogy (see Chapter 3). Though I often refer to them in the singular, each is really a family of arguments united by a common schema, a logical form with placeholders for premises that can be filled in a variety of ways. So to be more precise in this review, I'll refer to the *arguments from order*, etc.

The arguments by analogy assert that, given all the ways in which the universe resembles a machine, we should conclude that it also resembles a machine in having a designer. This family of arguments was brought into its sharpest focus and then undermined by David Hume (see Chapter 7). The arguments from purpose appeal to natural phenomena, especially the ways in which living things are adapted to their environment, that are supposed to self-evidently exhibit purpose. In Chapter 10, we saw that we must disambiguate senses of 'purpose' to avoid begging the question. After all, what we are trying to establish is whether any portion of the natural universe was purposefully arranged by a designing intelligence. But when this is done the argument collapses – it is not the case that the sort of 'purpose' which can be observed in the natural world entails or even suggests design.

Arguments from order proceed by first establishing a special empirical property or set of properties that is always (or nearly always) the product of intelligent agency. Each then argues that the special property is in fact possessed by some natural object, and concludes that this natural object – possibly the universe as a whole – can be said to have a designer. Arguments from order are well represented in the modern literature. In particular, the arguments we examined in Chapters 12 and 13 that appeal to the structure of living things at the molecular scale are arguments from order. Finally, arguments from providence are those which begin by noting many ways in which the world is configured precisely as it must be for human beings to survive, flourish, learn, or achieve other ends of evident value. From this, each such argument concludes that the universe was contrived or arranged for the benefit of humans, and thus must have a designer. This general strategy is very much alive in the twenty-first century, and is probably the most philosophically important type of design argument of the modern era. In particular, the contentious fine-tuning arguments are clear instances of modern arguments from providence.

Lumping design arguments into only four types masks and omits a great deal of diversity. For starters, there are many different arguments belonging to each of the four categories listed above, as we discovered in our historical survey. Furthermore, the four major types do not even capture all forms of design argument. For instance, Berkeley's divine-language argument or Reid's argument from direct intuition do not fall into any of these categories (see Chapter 6). And while I have tried to capture all of the major forms of design argument, there are many specific arguments and variations of arguments that could not be fitted into a single book of this length. Of course, cataloguing the diversity of arguments was only a means to another end. The central question that concerns us is whether any of the contenders are sound or cogent. Is there a convincing design argument?

19.2 So what's the answer?

The short answer is that none of the design arguments considered in this book is entirely compelling. Each is lacking in one respect or another. But the differences are important. Some are irredeemable failures for reasons of logical structure. For example, Aquinas' Fifth Way commits a quantifier-shift fallacy by inferring 'there exists a y such that for all x' from 'for all x there exists a y'. The argument from purpose, insofar as it is distinct from the argument from order, is either viciously circular or question-begging.

And various definitions of Dembski's specified complexity are internally inconsistent and inconsistent with one another.

Other arguments, however, are supported by a perfectly acceptable logical structure. There are instead problems with the premises on which they rest. For instance, some versions of the argument from order assume that there are only two possible causes or explanations for the adaptedness of living things: chance and mind. The importance of Darwin's theory of evolution for design arguments is that the existence of such a theory falsifies this premise by demonstrating a third possibility. But the argument from order can be viewed as a (rather trivial) argument template awaiting the identification of an appropriate premise – in this case, awaiting the identification of a physical property that can do the job of the placeholder term, 'order'. What would such a premise and its supporting argument look like? Well, we believe on thermodynamic grounds that it is impossible for certain chemical reactions to occur spontaneously, that is, without an input of energy from an external source. One can imagine having a theory of intelligence that entails a similar law: a certain property ('order' is the generic term I've been using) does not occur without the intervention of something intelligent. Of course, the thermodynamic claim rests on a complex theory that is in turn supported by a huge range of empirical evidence claims. Similarly, we would have to develop a clear characterization of intelligence and support a concomitant theory of which processes can and cannot result from intelligence. I will not speculate on the prospects of producing such a theory of intelligence, only stress that it is at least logically possible.

Finally, there are those arguments, particularly fine-tuning arguments, for which the grounds for rejection are less certain. Whether fine-tuning arguments are cogent depends on how we clear up ambiguities concerning the phenomena to be explained, what we take to count as a satisfying explanation, and how strong a metaphysical picture of natural laws we accept or reject. The verdict on these design arguments thus turns on debates at the heart of modern analytic metaphysics and the philosophy of science. What's a theory? What's an explanation, and when can we reasonably demand one? Does it make sense to speak of selection processes on universes? What laws are possible? What is the relevant sense of possibility? Until these questions are answered, it is impossible to pass firm judgment on the fine-tuning arguments.

19.3 The philosophical value of design arguments

Even if they are all failures, design arguments have been and, I suggest, continue to be of great scientific and philosophical value. In the early days of natural science, design arguments motivated the collection of important facts about natural history and provided a clear perspective from which to articulate alternate views of empirical inference. It was in large measure by pushing against and seeking alternatives to the very appealing use of design to explain the natural world that philosophers, natural philosophers, and eventually professional scientists were able to articulate a new empirical epistemology. By arguing against various appeals to divine agency, they constructed new norms for the explanation of natural phenomena.

The allure of producing a sound design argument has driven conceptual innovation important for both modern science and philosophy. The arguments we have looked at have driven much work on theories of complexity, on the likelihood approach to inductive inference, on the separation of teleology and agency, and on the interpretation of probability. Philosophically, design arguments have served as proving grounds for important theses. For instance, Berkeley's divine-language argument rests squarely on Descartes' test for intelligence. Berkeley, of course, did not think he was offering a *reductio ad absurdum*. But one person's modus ponens is another's modus tollens, and the modern reader is likely to view Berkeley's argument as an indictment of Descartes' position. Either way, this is clearly an instance in which a design argument put a philosophical thesis to the test by extracting surprising consequences relevant to the question of design. In the modern debate over fine tuning, design arguments are again forcing us to clarify muddled positions on the character of natural laws, the relation of physical theory to evidence, concepts of probability, and the boundary between physics and metaphysics. These are all important philosophical tasks. Whatever the ultimate verdict on design, the debate over design arguments will leave future generations with a much richer, much clearer view of empirical science and the reach of a posteriori argument. I envy the future scholar who sits down to write a book like this one in another 2,000 years.

19.4 The conversation continues

Arguments from the fine tuning of physical laws are a long way from the orrery argument of Cicero. This is not surprising considering the two millennia of empirical study and philosophical debate that have transpired since the heyday of Stoicism. What is remarkable, however, is that there are still new arguments to be made. After 2,000 years of sustained debate in the West, there are still novel ways to argue for design from the features of the world as we find it. I suggest that it is worthwhile continuing to debate how in principle a successful design argument could be made and whether any are already available. One reason this debate is fruitful is that it helps us clarify and rationally manage a persistent intuition. I said at the outset of this book that one of the ways we human beings make sense of the world is by treating it like one of us: We explain by attributing willful design and intentional action to features of nature, just as we explain the actions and artifacts of our fellow humans. To do so seems to be an unavoidable aspect of our nature as reasoners. Since the intuition is not likely to vanish anytime soon, a healthy philosophical debate is needed to keep it honest.

Bibliography

Anon. 1819. *Dialogues on Entomology, in Which the Forms and Habits of Insects Are Familiarly Explained.* London: R. Hunter.

　　1833. "Bernard Nieuwentyt, the Real Author of Paley's 'Natural Theology'." In *The Book of Days: A Miscellany of Popular Antiquities in Connection with the Calendar,* ed. R. Chambers, 196–97. London: W. & R. Chambers.

　　1848. "Dr. Paley's 'Natural Theology'." *The Athenæum,* September 9: 907–8.

　　1849. "The Charge of Plagiarism against Dr. Paley." *Methodist Quarterly Review* 31: 159–61.

　　1911. "Paley, William." In *The Encyclopædia Britannica.* Cambridge University Press.

Abbott, Cyril E. 1934. "How *Megarhyssa* Deposits Her Eggs." *Journal of the New York Entomological Society* 42, no. 1: 127–33.

Aertsen, Jan A. 1993. "Aquinas's Philosophy in Its Historical Setting." In *The Cambridge Companion to Aquinas,* ed. Norman Kretzmann and Eleonore Stump, 12–37. Cambridge University Press.

Aquinas, Thomas. 1920. *The "Summa theologica" of St. Thomas Aquinas,* 10 vols., trans. Fathers of the English Dominican Province. Vol. I, Part I, Qq. I–XXVI. London: Burns Oates & Washbourne.

Ariew, Roger. 2008. "Pierre Duhem." In *The Stanford Encyclopedia of Philosophy,* ed. Edward N. Zalta (fall 2008 edn.). http://plato.stanford.edu/archives/fall2008/entries/duhem. Accessed March 10, 2009.

Audi, Robert, ed. 1999. *The Cambridge Dictionary of Philosophy,* 2nd edn. Cambridge University Press.

Baker, Alan. 2011. "Simplicity." In *The Stanford Encyclopedia of Philosophy,* ed. Edward N. Zalta (summer 2011 edn.). http://plato.stanford.edu/archives/sum2011/entries/simplicity. Accessed September 20, 2013.

Baltzly, Dirk. 2008. "Stoicism." In *The Stanford Encyclopedia of Philosophy,* ed. Edward N. Zalta (fall 2008 edn.). http://plato.stanford.edu/archives/fall2008/entries/stoicism. Accessed December 30, 2008.

Barrow, John D., and Frank J. Tipler. 1986. *The Anthropic Cosmological Principle.* Oxford University Press.

Behe, Michael. 1996. *Darwin's Black Box: The Biochemical Challenge to Evolution.* New York: The Free Press.

2003. "The Modern Intelligent Design Hypothesis." In *God and Design*, ed. Neil Manson, 277–91. New York: Routledge.

Bennett, Charles H. 1995. "Logical Depth and Physical Complexity." In *The Universal Turing Machine: A Half-Century Survey*, ed. Rolf Herken, 207–35. New York: Springer.

Bentley, Richard. 1966. *Eight Sermons Preached at the Hon. Robert Boyle's Lecture, in the Year MDCXCII.* In *The Works of Richard Bentley, D.D.*, ed. Alexander Dyce. New York: AMS Press.

Benton, Michael J. 2010. "Studying Function and Behavior in the Fossil Record." *PLoS Biology* 8, no. 3: e1000321.

Berkeley, George. 1803. *Alciphron, or the Minute Philosopher.* New Haven, CT: Increase Cooke & Co.

1901. *Alciphron; or, the Minute Philosopher.* In *The Works of George Berkeley*, ed. Alexander Campbell Fraser, vol. II. Oxford University Press.

Berman, David. 1988. *A History of Atheism in Britain: From Hobbes to Russell.* London: Croom Helm.

Blakey, Robert. 1859. *Old Faces in New Masks.* London: W. Kent & Co.

Boyle, Robert. 1688. *A Disquisition about the Final Causes of Natural Things.* London: John Taylor.

Burnell, S. Jocelyn Bell. 2004. "Little Green Men, White Dwarfs or Pulsars?" *Cosmic Search* 1, no. 1. www.bigear.org/vol1no1/burnell.htm. Accessed September 16, 2013.

Cairns-Smith, A. G. 1985. *Seven Clues to the Origin of Life.* Cambridge University Press.

Carr, B. J., and M. J. Rees. 1979. "The Anthropic Principle and the Structure of the Physical World." *Nature* 278: 605–12.

Chaitin, Gregory J. 1966. "On the Length of Programs for Computing Finite Binary Sequences." *Journal of the ACM* 13, no. 4: 547–69.

1975a. "Randomness and Mathematical Proof." *Scientific American* 232, no. 5: 47–52.

1975b. "A Theory of Program Size Formally Identical to Information Theory." *Journal of the ACM* 22, no. 3: 329–40.

1977. "Algorithmic Information Theory." *IBM Journal of Research and Development* 21: 350–59.

1979. "Toward a Mathematical Definition of 'Life'." In *The Maximum Entropy Formalism: A Conference Held at the Massachusetts Institute of Technology on May*

2–4, 1978, ed. R. D. Levine and M. Tribus, 477–98. Cambridge, MA: MIT Press.

Cicero. 1998. *The Nature of the Gods*, trans. P. G. Walsh, Oxford World's Classics. Oxford University Press.

Cohen, J. Bernard. 1985. *Revolution in Science*. Cambridge, MA: The Belknap Press of Harvard University Press.

Collins, Robin. 2003. "Evidence for Fine-Tuning." In *God and Design*, ed. Neil Manson, 178–99. New York: Routledge.

2009. "The Teleological Argument: An Exploration of the Fine-Tuning of the Universe." In *The Blackwell Companion to Natural Theology*, ed. William Lane Craig and James Porter Moreland, 202–81. Chichester, UK and Malden, MA: Wiley-Blackwell.

Copleston, Frederick. 1993. *A History of Philosophy*, 9 vols. Vols. I and V. New York: Image Books.

Cornelis, Guy R. 2006. "The Type III Secretion Injectisome." *Nature Reviews Microbiology* 4: 811–25.

Craig, William Lane. 2003. "Design and the Anthropic Fine-Tuning of the Universe." In *God and Design*, ed. Neil Manson, 155–77. New York: Routledge.

Cummins, Robert. 1975. "Functional Analysis." *Journal of Philosophy* 72, no. 20: 741–65.

Curtis, Helena, and N. Sue Barnes. 1989. *Biology*, 5th edn. New York: Worth Publishers.

Darwin, Charles. 1838–51. "'Books to Be Read' and 'Books Read' Notebook." CUL-DAR119. Transcribed by Kees Rookmaaker. Darwin Online. http://darwin-online.org.uk. Accessed September 13, 2013.

1902. *Charles Darwin: His Life Told in an Autobiographical Chapter, and in a Selected Series of His Published Letters*, ed. Francis Darwin, 2nd edn. London: John Murray.

1993. *The Origin of Species*, 6th edn. New York: The Modern Library.

Davidson, Herbert A. 1987. *Proofs for Eternity, Creation and the Existence of God in Medieval Islamic and Jewish Philosophy*. Oxford University Press.

Davies, P. C. W. 1992. *The Mind of God: The Scientific Basis for a Rational World*. New York: Simon & Schuster.

2003. "The Appearance of Design in Physics and Cosmology." In *God and Design*, ed. Neil Manson, 147–54. New York: Routledge.

2007. *Cosmic Jackpot: Why Our Universe Is Just Right for Life*. Boston: Houghton Mifflin.

Dean, Jeffrey, Daniel Aneshansley, Harold Edgerton, and Thomas Eisner. 1990. "Defensive Spray of the Bombardier Beetle: A Biological Pulse Jet." *Science* 248, no. 4960: 1219–21.

Delfgaauw, Bernard. 1968. *The Student History of Philosophy*, trans. B. D. Smith. Albany, NY: Magi Books.

Dembski, William A. 2002. *No Free Lunch: Why Specified Complexity Cannot Be Purchased without Intelligence*. Lanham, MD: Rowman & Littlefield.

2004. "The Logical Underpinnings of Intelligent Design." In *Debating Design: From Darwin to DNA*, ed. William Dembski and Michael Ruse, 311–30. Cambridge University Press.

2005. "Specification: The Pattern That Signifies Intelligence." www. designinference.com/documents/2005.06.Specification.pdf. Accessed September 10, 2013.

2012. "Design Inference Website." www.designinference.com. Accessed August 6, 2012.

Dembski, William A., and Jonathan Wells. 2008. *The Design of Life: Discovering Signs of Intelligence in Biological Systems*. Dallas: Foundation for Thought and Ethics.

Denton, Michael J. 1998. *Nature's Destiny: How the Laws of Biology Reveal Purpose in the Universe*. New York: The Free Press.

Denton, Michael, Craig Marshall, and Michael Legge. 2001. "The Protein Folds as Platonic Forms: New Support for the Pre-Darwinian Conception of Evolution by Natural Law." *Journal of Theoretical Biology* 219: 325–42.

Derham, William. 1760. *Select Remains of the Learned John Ray, M.A. and F.R.S. with His Life*. London: George Scott.

1786a. *Derham's Physico and Astro Theology; or, a Demonstration of the Being and Attributes of God*, 2 vols. Vol. I. London: J. Walter.

1786b. *Derham's Physico and Astro Theology; or, a Demonstration of the Being and Attributes of God*, 2 vols. Vol. II. London: J. Walter.

Descartes, René. 2006. *A Discourse on the Method of Correctly Conducting One's Reason and Seeking Truth in the Sciences*, trans. Ian Maclean. Oxford University Press.

Dryden, John, and A. H. Clough. 1876. *Plutarch's Lives of Illustrious Men*. Boston: Little, Brown, and Company.

Dutton, Denis. 2009. *The Art Instinct: Beauty, Pleasure, and Human Evolution*. New York: Bloomsbury Press.

Edwards, A. W. F. 1972. *Likelihood*. Cambridge University Press.

Einstein, Albert. 1961. *Relativity*. New York: Crown Trade Paperbacks.

Eisner, Thomas. 2003. *For Love of Insects*. Cambridge, MA: The Belknap Press of Harvard University Press.

Esposito, John, ed. 1999. *The Oxford History of Islam*. Oxford University Press.

Falcon, Andrea. 2012. "Aristotle on Causality." In *The Stanford Encyclopedia of Philosophy*, ed. Edward N. Zalta (winter 2012 edn.). http://plato.stanford.edu/archives/win2012/entries/aristotle-causality. Accessed September 20, 2013.

Fee, Jerome. 1941. "Maupertuis, and the Principle of Least Action." *Scientific Monthly* 52, no. 6: 496–503.

Feynman, Richard. 1985. *QED: The Strange Theory of Light and Matter*. Princeton University Press.

Frade, Miguel, F. Fernandez de Vega, and Carlos Cotta. 2008. "Modelling Video Games' Landscapes by Means of Genetic Terrain Programming – a New Approach for Improving Users' Experience." In *Applications of Evolutionary Computing*, ed. Mario Giacobini, Anthony Brabazon, Stefano Cagnoni, Gianni A. Di Caro, Rolf Drechsler, Anikó Ekárt, and Anna I. Esparcia-Alcázar, 485–90. Berlin: Springer.

Fraser, Alexander Campbell, ed. 1911. *Selections from Berkeley*, 6th edn. Oxford: Clarendon Press.

Freeth, T., Y. Bitsakis, X. Moussas, J. H. Seiradakis, A. Tselikas, H. Mangou, M. Zafeiropoulou, R. Hadland, D. Bate, A. Ramsey, M. Allen, A. Crawley, P. Hockley, T. Malzbender, D. Gelb, W. Ambrisco, and M. G. Edmunds. 2006. "Decoding the Ancient Greek Astronomical Calculator Known as the Antikythera Mechanism." *Nature* 444: 587–91.

Gardner, Martin. 1986. "WAP, SAP, PAP, & FAP." *New York Review of Books* 33, no. 8: 22–25.

Gell-Mann, Murray. 1995. "What Is Complexity?" *Complexity* 1, no. 1: 16–19.

Gell-Mann, Murray, and Seth Lloyd. 2004. "Effective Complexity." In *Nonextensive Entropy: Interdisciplinary Applications*, ed. Murray Gell-Mann and Constantino Tsallis, 387–98. Oxford University Press.

Gerson, L. P. 1994. *God and Greek Philosophy*. London: Routledge.

Gillespie, Neal C. 1990. "Divine Design and the Industrial Revolution: William Paley's Abortive Reform of Natural Theology." *Isis* 81, no. 2: 214–29.

Glass, Marvin, and Julian Wolfe. 1986. "Paley's Design Argument for God." *Sophia* 25, no. 2: 17–19.

Godfrey-Smith, Peter. 1993. "Functions: Consensus without Unity." *Pacific Philosophical Quarterly* 74: 196–208.

Grassberger, Peter. 1986. "Toward a Quantitative Theory of Self-Generated Complexity." *International Journal of Theoretical Physics* 25, no. 9: 907–38.

Hacking, Ian. 1987. "The Inverse Gambler's Fallacy: The Argument from Design. The Anthropic Principle Applied to Wheeler Universes." *Mind* 96, no. 383: 331–40.

Halliday, David, Robert Resnick, and Jearl Walker. 1993. *Fundamentals of Physics*, 4th edn. New York: John Wiley & Sons.

Hansson, Sven Ove. 2008. "Science and Pseudo-Science." In *The Stanford Encyclopedia of Philosophy*, ed. Edward N. Zalta (fall 2008 edn.). http://plato. stanford.edu/archives/fall2008/entries/pseudo-science. Accessed March 10, 2009.

Harman, Gilbert. 1965. "The Inference to the Best Explanation." *Philosophical Review* 74, no. 1: 88–95.

Harrell, Mara. 2010. "Creating Argument Diagrams." www.hss.cmu.edu/ philosophy/harrell/Creating_Argument_Diagrams.pdf. Accessed August 28, 2010.

Hawkins, Roger. 2012. "Robert Blakey, Part 1." Morpeth Antiquarian Society, www.northumbriana.org.uk/antiquarian%20society/blakey1.htm. Accessed December 30, 2012.

Herschel, John F. W. 1831. *A Preliminary Discourse on the Study of Natural Philosophy*, ed. Dionysius Lardner, The Cabinet Cyclopaedia. London: Longman, Rees, Orme, Brown, and Green.

Hesse, Mary B. 1964. "Models and Matter." In *Quanta and Reality: A Symposium for the Non-Scientist on the Physical and Philosophical Implications of Quantum Mechanics*, ed. British Broadcasting Corporation, 49–57. Cleveland: Meridian Books.

Hick, John. 1963. *Philosophy of Religion*. Englewood Cliffs, NJ: Prentice-Hall.

Hicks, L. E. 1883. *A Critique of Design-Arguments*. New York: Charles Scribner's Sons.

Hoyle, Fred. 1982. "The Universe: Past and Present Reflections." *Annual Review of Astronomy and Astrophysics* 20, no. 1: 1–36.

  1983. *The Intelligent Universe*. New York: Holt, Reinhart and Winston.

Hume, David. 1777. *The Life of David Hume*. London: W. Strahan and T. Cadell.

  1990. *Dialogues Concerning Natural Religion*, ed. Martin Bell. London: Penguin Books.

Hurlbutt, Robert H. 1965. *Hume, Newton, and the Design Argument*. Lincoln, NE: University of Nebraska Press.

Hutcheson, Francis. 1726. *An Inquiry into the Original of Our Ideas of Beauty and Virtue*, 2nd edn. London: Printed for J. Darby, A. Bettesworth, F. Fayram, J. Pemberton, C. Rivington, J. Hooke, F. Clay, J. Batley, and E. Symon.

Hyman, Arthur, and James J. Walsh, eds. 1973. *Philosophy in the Middle Ages*, 2nd edn. Indianapolis: Hackett Publishing Company.

Janet, Paul. 1884. *Final Causes*, 2nd edn., trans. Robert Flint. New York: Charles Scribner's Sons.

Jourdain, Philip E. B. 1913. *The Principle of Least Action*. Chicago: Open Court.

Juhl, Cory. 2006. "Fine-Tuning Is Not Surprising." *Analysis* 66, no. 4: 269–75.

Kenny, Anthony. 1969. *The Five Ways: St. Thomas Aquinas' Proofs of God's Existence*. New York: Schocken Books.

Klee, Robert. 2002. "The Revenge of Pythagoras: How a Mathematical Sharp Practice Undermines the Contemporary Design Argument in Astrophysical Cosmology." *British Journal for the Philosophy of Science* 53, no. 3: 331–54.

Kline, A. David. 1993. "Berkeley's Divine Language Argument." In *George Berkeley, "Alciphron; or, the Minute Philosopher": In Focus*, ed. David Berman, 185–99. London and New York: Routledge.

Kolmogorov, Andrei. 1968. "Logical Basis for Information Theory and Probability Theory." *IEEE Transactions of Information Theory* IT-14, no. 5: 662–64.

Kretzmann, Norman, and Eleonore Stump, eds. 1993. *The Cambridge Companion to Aquinas*. Cambridge University Press.

Kühner, Raphael. 1847. *Xenophon's Memorabilia of Sokrates*, trans. George B. Wheeler. London: William Allan.

La Salle, John, Mohsen Ramadan, and Bernarr R. Kumashiro. 2009. "A New Parasitoid Wasp of the Erythrina Gall Wasp, *Quadrastichus erythrinae* Kim (Hymenoptera: Eulophidae)." *Zootaxa* 2083: 19–26.

Lacey, A. R. 1996. *A Dictionary of Philosophy*, 3rd edn. New York: Barnes and Noble.

Langer, William L. 1948. *An Encyclopedia of World History*. Boston: Houghton Mifflin Company.

Lanham, Url. 1964. *The Insects*. New York: Columbia University Press.

Lempel, Abraham, and Jacob Ziv. 1976. "On the Complexity of Finite Sequences." *IEEE Transactions of Information Theory* IT-22, no. 1: 75–81.

Lennox, James G. 2001. *Aristotle's Philosophy of Biology: Studies in the Origins of Life Science*. Cambridge University Press.

Leslie, John. 1989. *Universes*. London and New York: Routledge.

Livio, M., D. Hollowell, A. Weiss, and J. W. Truran. 1989. "The Anthropic Significance of the Existence of an Excited State of 12C." *Nature* 340, no. 6231: 281–84.

Lloyd, Seth. 2001. "Measures of Complexity: A Non-Exhaustive List." *IEEE Control Systems Magazine*, August: 7–8.

    2002. "Computational Capacity of the Universe." *Physical Review Letters* 88, no. 23: 237901.

Lloyd, Seth, and Heinz Pagels. 1988. "Complexity as Thermodynamic Depth." *Annals of Physics* 188: 186–213.

McGrew, Timothy, Lydia McGrew, and Eric Vestrup. 2003. "Probabilities and the Fine-Tuning Argument: A Skeptical View." In *God and Design*, ed. Neil Manson, 200–8. New York: Routledge.

MacKenzie, Donald A. 1915. *Myths of Babylonia and Assyria*. London: The Gresham Publishing Company.

McKeon, Richard, ed. 1941. *The Basic Works of Aristotle*. New York: Random House.

Maclaurin, Colin. 1750. *An Account of Sir Isaac Newton's Philosophical Discoveries*, 2nd edn. London: A. Millar.

Malthus, Thomas. 1960. "A Summary View of the Principle of Population." In *Three Essays on Population*. New York: The New American Library.

Mandelbrot, Benoit B. 1983. *The Fractal Geometry of Nature*. New York: W. H. Freeman and Company.

Marden, James. 2003. "The Surface-Skimming Hypothesis for the Evolution of Insect Flight." *Acta zoologica cracoviensia* 46 (Supplement: Fossil Insects): 73–84.

Marion, Jerry B., and Stephen T. Thornton. 1995. *Classical Dynamics of Particles and Systems*, 4th edn. Fort Worth: Saunders College Publishing.

Mill, John Stuart. 1998. *Three Essays on Religion*. Amherst, NY: Prometheus Books.

Miller, Kenneth R. 2003. "Answering the Biochemical Argument from Design." In *God and Design*, ed. Neil Manson, 292–307. New York: Routledge.

Miller, Perry. 1978. "Bentley and Newton." In *Isaac Newton's Papers and Letters on Natural Philosophy and Related Documents*, ed. I. Bernard Cohen and Robert E. Schofield, 271–87. Cambridge, MA: Harvard University Press.

Mitchell, Melanie. 2009. *Complexity: A Guided Tour*. Oxford University Press.

Morgan, C. Lloyd. 1906. *The Interpretation of Nature*. New York: The Knickerbocker Press.

Munitz, Milton Karl. 1981. *Space, Time, and Creation: Philosophical Aspects of Scientific Cosmology*. New York: Dover Publications.

Newton, Isaac. 1995. *The Principia*, trans. Andrew Motte, Great Minds Series. Amherst, NY: Prometheus Books.

Nieuwentyt, Bernard. 1721. *The Religious Philosopher; or, the Right Use of Contemplating the Works of the Creator*, trans. John Chamberlayne, 2nd edn. London: J. Senex and W. Taylor.

Norton, John D. 2008. "Ignorance and Indifference." *Philosophy of Science* 75, no. 1: 45–68.

Oppy, Graham. 2002. "Paley's Argument for Design." *Philo* 5, no. 2: 161–73.

Paley, William. 1802. *Natural Theology*, 2nd edn. London: R. Faulder.

2006. *Natural Theology*, ed. Matthew Eddy and David Knight. Oxford University Press.

Parshall, Karen. 1982. "Varieties as Incipient Species: Darwin's Numerical Analysis." *Journal of the History of Biology* 15, no. 2: 191–214.

Pease, Arthur Stanley. 1941. "Caeli Enarrant." *Harvard Theological Review* 34, no. 3: 163–200.

Peirce, Charles Sanders. 1955. "Abduction and Induction." In *Philosophical Writings of Peirce*, ed. Justus Buchler, 150–56. Mineola, NY: Dover Publications.

Penrose, Roger. 2004. *The Road to Reality*. New York: Vintage Books.

Popkin, Richard H. 1992. "David Berman, *A History of Atheism in Britain: From Hobbes to Russell* (Book Review)." *Journal of the History of Philosophy* 30, no. 1: 143.

Popper, Karl. 2002. *Conjectures and Refutations*. New York: Routledge.

Price, Derek de Solla. 1974. "Gears from the Greeks: The Antikythera Mechanism – a Calendar Computer from *Ca.* 80 B.C." *Transactions of the American Philosophical Society* 64, no. 7: 3–70.

Psillos, Stathis. 2002. *Causation and Explanation*. Montreal: McGill-Queen's University Press.

Quicke, Donald, Paul Wyeth, James D. Fawke, Hasan H. Basibuyuk, and Julian F. V. Vincent. 1998. "Manganese and Zinc in the Ovipositors and Mandibles of Hymenopterous Insects." *Zoological Journal of the Linnean Society* 124: 387–96.

Quine, W. V. O. 1951. "Two Dogmas of Empiricism." *Philosophical Review* 60, no. 1: 20–43.

Ratzsch, Del. 2001. *Nature, Design, and Science*. Albany, NY: State University of New York Press.

2003. "Perceiving Design." In *God and Design*, ed. Neil A. Manson, 124–44. New York: Routledge.

2008. "Teleological Arguments for God's Existence." In *The Stanford Encyclopedia of Philosophy*, ed. Edward N. Zalta (fall 2008 edn.). http://plato.stanford.edu/archives/fall2008/entries/teleological-arguments. Accessed February 10, 2009.

Ray, John. 1714. *The Wisdom of God Manifested in the Works of the Creation: In Two Parts*, 6th edn. London: William Innys.

Reid, Thomas. 1785. *Essays on the Intellectual Powers of Man*. Edinburgh: John Bell.

1975. *Thomas Reid's "Inquiry" and "Essays,"* ed. Ronald E. Beanblossom and Keith Lehrer. Indianapolis: Bobbs-Merrill.

1981. *Thomas Reid's "Lectures on Natural Theology" (1780)*, ed. Elmer H. Duncan. Washington, DC: University Press of America.

Riley, C. V. 1888. "The Habits of *Thalessa* and *Tremex*." *Insect Life* 1, no. 6: 168–79.

Rissanen, J. 1978. "Modeling by Shortest Data Description." *Automatica* 14: 465–71.

Ruse, Michael. 1999. *The Darwinian Revolution*, 2nd edn. University of Chicago Press.

Sagan, Carl. 1979. *Broca's Brain: Reflections on the Romance of Science*. New York: Random House.

Sedley, David. 2007. *Creationism and Its Critics in Antiquity*. Berkeley: University of California Press.

Shackel, Nicholas. 2007. "Bertrand's Paradox and the Principle of Indifference." *Philosophy of Science* 74, no. 2: 150–75.

Shupbach, Jonah. 2005. "Paley's Inductive Inference to Design: A Response to Graham Oppy." *Philosophia Christi* 7, no. 2: 491–502.

Simon, Herbert A. 1962. "The Architecture of Complexity." *Proceedings of the American Philosophical Society* 106, no. 6: 467–82.

Sinnigen, William G., and Arthur E. R. Boak. 1977. *A History of Rome to A.D. 565*. New York: Macmillan.

Smolin, Lee. 1997. *The Life of the Cosmos*. Oxford University Press.

Snyder, Laura J. 2009. "William Whewell." In *The Stanford Encyclopedia of Philosophy*, ed. Edward N. Zalta (winter 2009 edn.). http://plato.stanford.edu/archives/win2009/entries/whewell. Accessed September 20, 2013.

Sober, Elliott. 2000. *Philosophy of Biology*, 2nd edn. Boulder, CO: Westview Press.
  2003. "The Design Argument." In *God and Design*, ed. Neil Manson, 27–54. New York: Routledge.
  2005a. *Core Questions in Philosophy*, 4th edn. Upper Saddle River, NJ: Prentice Hall.
  2005b. "The Design Argument." In *The Blackwell Guide to the Philosophy of Religion*, ed. William E. Mann, 117–47. Malden, MA: Blackwell Publishing.
  2008. *Evidence and Evolution: The Logic behind the Science*. Cambridge University Press.

Solomonoff, Ray J. 1964a. "A Formal Theory of Inductive Inference, Part I." *Information and Control* 7, no. 1: 1–22.
  1964b. "A Formal Theory of Inductive Inference, Part II." *Information and Control* 7, no. 2: 224–54.

Steering Committee on Science and Creationism, National Academy of Sciences. 1999. *Science and Creationism: A View from the National Academy of Sciences*, 2nd edn. Washington, DC: National Academies Press.

Stenger, Victor J. 2011. *The Fallacy of Fine-Tuning: Why the Universe Is Not Designed for Us*. Amherst, NY: Prometheus Books.

Stone, Graham N., and James M. Cook. 1998. "The Structure of Cynipid Oak Galls: Patterns in the Evolution of an Extended Phenotype." *Proceedings of the Royal Society London B* 265: 979–88.

Swartz, Norman. 2009. "Laws of Nature." www.iep.utm.edu/lawofnat. Accessed May 24, 2013.

Terence [Publius Terentius Afer]. 1886. *Adelphi* (The brothers), ed. A. Sloman. Oxford: Clarendon Press.

Todhunter, Isaac. 1865. *A History of the Mathematical Theory of Probability.* Cambridge: Macmillan and Co.

Valentine, M. 1885. *Natural Theology; or, Rational Theism*, 9th edn. Boston: Silver, Burdett & Co.

van Fraassen, Bas C. 1989. *Laws and Symmetry.* Oxford University Press.

"Verax." 1848. "Dr. Paley's 'Natural Theology'." *The Athenæum*, August 12: 803.

Vickers, John. 2009. "The Problem of Induction." In *The Stanford Encyclopedia of Philosophy*, ed. Edward N. Zalta (spring 2009 edn.). http://plato.stanford.edu/archives/spr2009/entries/induction-problem. Accessed September 20, 2013.

von Mises, Richard. 1981. *Probability, Statistics, and Truth.* New York: Dover Publications.

Waltz, George H., Jr. 1950. "Fooling the Jets for More Speed." *Popular Science*, April: 136–38.

Ward, Keith. 2003. "Teleology." In *Encyclopedia of Science and Religion*, ed. J. Wentzel Vrede van Huyssteen, 876–80. New York: Macmillan Reference USA.

Warn, Faith. 2000. *Bitter Sea: The Real Story of Greek Sponge Diving.* South Woodham Ferrers, UK: Guardian Angel.

Weisheipl, J. 1983. *Friar Thomas d'Aquino. His Life, Thought, and Works.* Washington, DC: Catholic University of America Press.

Whewell, William. 1833. *Astronomy and General Physics Considered with Reference to Natural Theology.* London: William Pickering.

Wright, Larry. 1973. "Functions." *Philosophical Review* 82, no. 2: 139–68.

Index